The Computer in Contemporary Cartography

PROGRESS IN CONTEMPORARY CARTOGRAPHY

Editor
D. R. Fraser Taylor
*Carleton University,
Ottawa, Canada*

Editorial Board

M. Jacques Bertin,
*Laboratoire de Graphique,
Paris, France*

Mr. David P. Bickmore,
*Experimental Cartography Unit,
London, U.K.*

Mr. Frederick R. Broome,
*Bureau of the Census,
Washington, DC, U.S.A.*

Professor Joel L. Morrison,
*University of Wisconsin,
Madison, U.S.A.*

Dr. David Rhind,
*University of Durham,
Durham, U.K.*

Dr. Bengt Rystedt,
*Central Board for Real Estate Data,
Gävle, Sweden*

Professor K. A. Salichtev,
*Lomonosov Moscow State University,
Moscow, U.S.S.R.*

Professor Bogodar Winid,
*University of Warsaw,
Warsaw, Poland*

Progress in Contemporary Cartography

Volume I

The Computer in Contemporary Cartography

Edited by

D. R. Fraser Taylor

*Carleton University,
Ottawa, Canada*

JOHN WILEY & SONS
Chichester · New York · Brisbane · Toronto

Copyright © 1980 by John Wiley & Sons, Ltd.

All rights reserved.

No part of this book may be reproduced by any means, nor transmitted, nor translated into a machine language without the written permission of the publisher.

British Library Cataloguing in Publication Data:

The computer in contemporary cartography.—
  (Progress in contemporary cartography; Vol. 1).
    1. Cartography—Data processing
    I. Taylor, D. R. F. II. Series
526'.028'54044   GA102.4.E4   79-42727

ISBN 0 471 27699 5

Typeset by Computacomp (UK) Ltd, Fort William, Scotland, and printed in the United States of America.

Notes about the Authors

D. R. Fraser Taylor is Professor of Geography and International Affairs at Carleton University, Ottawa, Canada. He received his Ph.D. from the University of Edinburgh and a post-graduate degree in education from the University of London. His interest in computer-assisted cartography began in 1966 while attending a course at Harvard. Dr. Taylor is co-editor of *The Computer and Africa: Applications, Problems and Potential* (Praeger, 1977), and has published widely both on cartographic topics and third-world development problems. Dr. Taylor is President of the Canadian Cartographic Association and Canadian representative on the I.C.A. Commission on Education. He is also Chairman of the Cartographic Education Committee of the National Commission on Cartography in Canada.

Ray Boyle received his B.Sc. and Ph.D. from the University of Birmingham, England, in 1940 and 1943 respectively. These degrees were in physical chemistry and his research was of a war-time nature concerned with theoretical aspects of the sensitivity of explosives. During this period he developed an intense interest in the application of electronics, or 'radio' as it was then called. After a period as a Scientific Research Officer with the British Admiralty, he became Technical and Managing Director of a specialized instrumentation company in Glasgow, Scotland, called D-mac, Ltd. His interests in electronics and graphics created new advances and patents in automatic drafting tables connected to radar systems, and later, in the 1950s, to the first large digital computers. This was followed by the necessity for development of an 'anti-plotter' or digitizer. In 1960 came cooperation with the Oxford University Press, and in 1964 the 'Oxford' system of automated cartography was presented to the International Cartographic Association in Edinburgh, Scotland. In 1965 he immigrated to Canada to become Full Professor of Electrical Engineering at the University of Saskatchewan. His interests in automated cartography led immediately to the growth of a large research team, at first funded by the Canadian Hydrographic Service and later, over the next 12 years, by the National Research Council and Defense Research Board of Canada, as well as by the U.S. Naval Oceanographic Office, the U.S. Department of Agriculture, the U.S. Geological Survey, and many others. Many themes and reports have been produced by this team, relative to computer-assisted cartography and geographic information systems. Many areas have been investigated by its researchers and personally by Ray Boyle acting as consultant to various cartographic establishments.

D. P. Bickmore is a practising cartographer. After leaving university, he worked for twelve months with John Bartholomew of Edinburgh before being involved in the war, where he served in the Directorate of Military Survey and in S.E. Asia. After the war he joined the staff of the Clarendon Press, Oxford, who in 1947 embarked on the production of the Oxford Atlas (1951). Subsequently a Cartographic Department was set up under Bickmore's direction which pioneered new techniques (e.g. Oxford Plastic Relief mapping in 1954; Oxford System of Automatic Cartography 1963) as well as a range of new atlases leading to the Atlas of Britain 1964. In the mid-1960s the Royal Society became interested in the potential of the new technology in mapping and urged the Natural Environment Research Council to develop it. In 1967 the Experimental Cartography Unit was established under Bickmore's direction, and has acquired an international reputation in the field of digital mapping and the preparation of a series of scientifically interesting maps, atlases, and reports. In 1979 he retired from E.C.U. and became Senior Research Fellow in Cartography at the Royal College of Art.

Stein W. Bie received his B.A. (Hons.) in Agricultural Science from Oxford University, England in 1967, and his M.A. and D.Phil. from the same university in 1973, on studies on quality control in soil and land resource surveys; he also holds a degree in Nutrition from Cambridge University. From 1972 to 1978 he worked for the Dutch Government creating a computer-based information system for the Netherlands Soil Survey Institute and the Netherlands Geological Survey. In 1978 he returned to his homeland, Norway, where he develops systems for natural resource data using statistics and computers at the Norwegian Computing Center. Bie has worked in the field in East Africa, Australia, Cyprus and the Netherlands, and holds offices in international organizations for soil science and geology. He has published some forty scientific papers on soil and land resource mapping, and edited a number of books on soil and geological data processing. His current interests are in the field of microprocessors for computer-aided cartography and in technology transfer to the developing world.

Frederick R. Broome is Chief of the Computer Graphics Staff, Geography Division, U.S. Bureau of the Census. He received a B.S. in Physical Geography from the University of Georgia and an M.A. in Economic Geography from the same university. Mr. Broome is a member of the Federal Mapping Task Force on Cartographic Activities in the U.S. Government and is also a member of A.C.S.M. and the C.C.A.

Christopher M. Gold is currently a Post-Doctoral Fellow in the Geology Department, University of Alberta, Canada, from which he received his Ph.D. in 1978. His academic interests include Quaternary geology, geological data processing and automated cartography, and laboratory instrumentation. He is currently Chairman of the Automation Interest Group of the Canadian Cartographic

NOTES ABOUT THE AUTHORS

Association and is involved in consulting and conducting workshops in automated cartography.

Brigadier Lewis J. Harris was educated at Christ College, Brecon; the Royal Military Academy, Woolwich and Pembroke College, Cambridge (Mechanical Science Tripos); commissioned in the Corps of Royal Engineers, he has been engaged on surveys and mapping for over 40 years since his first work on the Geodetic Triangulation of Jamaica, West Indies, 1937–39. During this period he was closely connected with the great changes in the practices of geodesy and topographical cartography caused by the application of advances in electronic technology, remote sensors, photogrammetry, computer graphics and cartographic depiction. He has been Director of Surveys, Middle East and East Africa; Director of the Map Production, Ordnance Survey; and Director of Military Survey, Ministry of Defence, U.K. He was a founding Vice-President of the International Cartographic Association, the first Chairman of the Cartographic Subcommittee of the Royal Society, and Honorary Foreign Secretary of the Royal Geographical Society. Since 1967, he has been a consultant to the Canadian Federal Surveys and Mapping Branch, and has been much concerned with computerized cartographic mapping.

Joel L. Morrison is Professor of Geography at the University of Wisconsin, Madison, U.S.A. Currently he serves as Department Chairman and is engaged in teaching and research in cartography. He received his B.A. degree from Miami University, Oxford, Ohio, and his M.S. and Ph.D. degrees from the University of Wisconsin, Madison. Dr. Morrison is the author of numerous articles in cartography; is co-author, with A. H. Robinson and R. D. Sale of the textbook *Elements of Cartography* and is an associate editor of *Goode's World Atlas*. He is a member of the American Congress on Surveying and Mapping, the Association of American Geographers, the British Cartographic Society, the Canadian Cartographic Association and the American Society of Photogrammetry. Dr. Morrison is currently Chairman of the Committee on State Cartography in Wisconsin. He is a past Chairman of the Cartography Division of A.C.S.M. and represents the United States on the International Cartographic Association's Commission on Cartographic Communication. He currently serves as Vice-President of A.C.S.M.

Lars Ottoson graduated from the Royal Institute of Technology in 1955. Until 1958 he worked as a research engineer at the division of photogrammetry of this institute with theoretical photogrammetric problems. In 1958 he joined the State Power Board where he was responsible for geodetic and photogrammetric activities in planning water-power stations. Then in 1963 he joined the Geographical Survey Office which in 1974 was reorganized to the National Land Survey. He now serves as head of the photogrammetric division of this agency and holds the responsibility for development of new techniques in photogrammetry and cartography.

David Rhind graduated from the University of Bristol in 1965 as a geographer, having given up the idea of becoming a physicist on the first day. He then proceeded to a Ph.D. in the University of Edinburgh, based on fluvioglacial landforms and, in 1969, carried out an early feasibility project on the data banking of drift borehole data. A then-record stay of 4 years in the Experimental Cartography Unit as Head of the Applications Section was followed by appointment as a lecturer at the University of Durham. He was promoted to a personal Readership in January 1978. In later years he has written widely on matters related to cartography and geographical data processing, has held about a dozen sizeable research grants and has served as Advisor to the Ordnance Survey Review Committee, consultant to the Department of Environment, etc.

Bengt Rystedt studied mathematics, statistics and computer science at the University of Lund. From 1965 to 1972 he was a researcher at the Departments of Geography and Building Function Analysis. In 1973 he received a doctoral degree with a thesis on computer cartography. From 1972 he has been responsible for the development work at the Central Board for Real Estate Data, a state agency under the Ministry of Justice.

Thomas C. Waugh received a B.Sc. (Hons.) in Geography and an M.Phil in Geography from the University of Edinburgh. From 1971 to 1978 he was a Senior Computing Officer and Honorary Fellow of Social Sciences at the Program Library Unit, University of Edinburgh. Currently, Mr. Waugh is an independent consultant, Director of GIMMS Ltd. and Research Associate in the Cartographic Research Unit, Carleton University, Ottawa. Mr. Waugh is the originator of the GIMMS program and is active in research, development and publication on computer-assisted cartography. He is a member of the Canadian Cartographic Association, the International Segment-Oriented Referencing System Association and has participated in several international seminars.

Sidney W. Witiuk is Coordinator, Geocartographics Group, Systems and Data Processing Branch, Statistics Canada. He received a B.Sc. in Mathematics from the University of British Columbia, an M.A. in Geography from Simon Fraser University and an M.Sc. in Computer Science from the University of British Columbia. He is past Chairman of the Automation Special Interest Group of the Canadian Cartographic Association, co-founder of the National Capital Geographic Information Processing Group, member of the executive of the International Segment-Oriented Referencing System Association, and a member of A.C.S.M., C.C.A., S.O.R.S.A. and N.C.G.I.P.G.

Contents

Preface xi
D. R. Fraser Taylor

1 **The Computer in Contemporary Cartography: Introduction** 1
D. R. Fraser Taylor

2 **Computer Technology and Cartographic Change** 5
Joel L. Morrison

3 **The Nature of Computer-Assisted Cartography** 25
David Rhind

4 **Development in Equipment and Techniques** 39
A. R. Boyle

5 **The Application of Computer Technology to Topographical Cartography** 59
Lewis J. Harris

6 **Computer-Assisted Cartography: Research and Applications in Sweden** 93
Lars Ottoson and Bengt Rystedt

7 **Computer-Assisted Soil Mapping** 123
Stein W. Bie

8 **Geological Mapping by Computer** 151
Christopher Gold

9 **Census Mapping by Computer** 191
Frederick R. Broome and Sidney W. Witiuk

10 **The Development of the GIMMS Computer Mapping System** 219
Thomas C. Waugh

11 **Future Research and Development in Computer-Assisted Cartography** 235
D. P. Bickmore

Subject Index 251

Preface

This volume is the first of a series on Progress in Contemporary Cartography and many people have helped in its production. I should like especially to acknowledge the advice of the editorial board and the support of Carleton University. Barbara George deserves a special vote of thanks for editorial assistance and typing, and Christine Earl for help with the figures. The volume would not have been possible without the enthusiastic support of an international group of contributors and I hope that I have not strained personal friendships too far by excessive editorial demands. Any remaining shortcomings are the sole responsibility of the editor.

Ottawa, July 1979 D. R. F. TAYLOR

List of Figures

5.1	Hierarchical structure of data items.	75
5.2	On-line computer-assisted cartographic system.	86
6.1	Large-scale mapping of property boundaries from local survey.	95
6.2	Large-scale map scribed on a Kongsberg flatbed plotter.	97
6.3	Decca lanes plotted and numbered by scribing on a Kongsberg plotter.	100
6.4	Place names for a 1/250,000 map.	102
6.5	Three grid points in a regular grid.	113
6.6	A reproduction in black and white of a contour map produced by the colour plotter.	114
6.7	A square grid net map produced at C.F.D.	115
6.8	A dot map showing an enlargement of an area of Figure 6.7.	116
6.9	Average personal income in Uppsala County.	117
6.10A	Weather observation in symbol form at 10:00 a.m., November 1, 1977.	118
6.10B	Enlargement of a single weather observation plotted by an electrostatic plotter.	119
6.11	Isarithmic maps produced by E.R.U.	120
7.1	Coordinatograph cross-slide digitizer.	126
7.2	Cross-slide digitizer with servo follower.	126
7.3	Mat or grid type digitizer.	127
7.4	The menu technique for text and symbol entry.	127
7.5	A combined digitizer–plotter.	129
7.6	A cathode ray tube used in monitoring digitizing.	130
7.7	Basic digitizing modes.	131
7.8	Soil map of part of Holland.	135
7.9	Ground water map of area shown in Figure 7.8.	136
7.10	Suitability of area for arable agriculture.	137
7.11	Suitability of area for recreation.	138
7.12	Typical hardware associated with a computer-assisted soils mapping system.	140
7.13	Isoline map of topographical variation produced on a line printer.	143
7.14	Isoline map of groundwater using barriers to alter interpolation.	143
7.15	Map of soil parent material produced on an electrostatic printer/plotter.	145

7.16	Soil map on a cell by cell basis	146
7.17	Allocation algorithm for drainage charges	146
7.18	Resulting drainage payment classification map	147
8.1	Wind River Basin, Wyoming, stratigraphic cross-section and block diagram	152
8.2	Part of William Smith's geological map of Britain, 1815	154
8.3	Distribution of magnetic survey data for Canada, 1955–73 and the resulting estimated magnetic declination for 1975	157
8.4	Regional distribution of nickel content in soil samples over the basement complex, Sierra Leone	161
8.4A	Cubic trend surface	161
8.4B	Moving average surface	161
8.5	Petit Lac (Lake Geneva) sediment samples	165
8.5A	Sampling locations	165
8.5B	Dendograph showing sample clusters	165
8.5C	Geochemical facies map	165
8.6	Normals to observed bedding orientation	168
8.6A	Raw data	168
8.6B	Grouped data	168
8.6C	Residuals from overall mean	168
8.7	Geological field data sheer	170
8.8	Sequence of geological events	172
8.8A	Rock types with older/younger relationships across boundaries	172
8.8B	Derivation of events sequence	172
8.9A	Stratigraphic summary graph of relationships between layer types 1–10	175
8.9B	Data structure representing some topographic features	175
8.10A	Profile looking down fold axis	180
8.10B	Intersection of geological contact with topographic surface	180
8.10C	Overburden ratio map	180
8.11	LEDA stratigraphic coding system	184
8.11A	Field data forms	184
8.11B	Application to a complex geological section	185
9.1	Elements of a typical computer-assisted cartography system	193
9.2	Flowchart of the geographically referenced data storage and retrieval system	195
9.3	A.M.F. Plot of Lemoyne urban area	197
9.4	Double-line plot of Lemoyne Urban Area	198
9.5	Different zoning patterns for a city area restructuring the data base for regional network analysis	199
9.6	Prototype field collection map—Kamloops CT-16	200
9.7	Isodemographic map	201
9.8	Computer-assisted production of index maps	202

LIST OF FIGURES

9.9	An example of the use of SYMAP	203
9.10	An example of the use of PREVU	204
9.11	Dasymmetrically weighted choropleth mapping	206
9.12	Geocartographic summaries of national level distributions	208
9.12A,B	U.S. Bureau of the Census, 1978	208
9.12C	Statistics Canada, 1978	209
9.13	Two examples of the use of GIMMS to plot census tract data	210
9.14	The Domestic Information Display System	212
9.15	Grid square map	213
10.1	A GIMMS plot on an inexpensive drum plotter	220
10.2	GIMMS symbolism chart	221
10.3	Types of text available in GIMMS	230

The Computer in Contemporary Cartography
Edited by D. R. F. Taylor
© 1980 John Wiley & Sons Ltd

Chapter 1
The Computer in Contemporary Cartography: Introduction

D. R. Fraser Taylor

This volume is the first of a series which will review and report significant progress in theories, methods, and empirical research in modern cartography. It is perhaps not surprising that the first volume deals with the impact of the computer on contemporary cartography because there is little doubt that the computer is influencing all aspects of the discipline and will continue to do so in an increasingly important way. Arguments about the role or potential role of the computer in cartography abound, but it is clear that no cartographer can afford to ignore the computer and its implications for our discipline. The main purpose of this volume is to present to advanced students, researchers and professionals interested in modern cartography a current review of theory, methodology, and research related to computer-assisted cartography.

Computer-assisted cartography has been chosen as a general term applying to all aspects of cartography where the computer is used as an aid. The International Cartographic Association (I.C.A.) originally used the term automated cartography and has produced a *Glossary of Automation Terms in Cartography* (I.C.A., 1976) but in the Foreword of that edition Edson makes an interesting comment:

> ... as soon as human actions become a decisive part of a system one can no longer speak of the system as automatic (or automated). Automation of cartography becomes completely impossible and automation in cartography should be identified as a particular process in the cartographic component of the thought model which can be performed automatically, as for example the scribing of a particular map image as described by digital data. Systems which perform such a process in the context of cartography are called 'Automated Cartographic Systems'. Such systems may work entirely without computers. In cases where a digital computer *is* used, but the human operators still have a function the system is called 'computer-assisted' or 'computer-aided'. This raises a question as to the appropriateness of the title 'Automation in Cartography' since the definitions are primarily oriented toward 'computer-assisted functions'. (I.C.A., 1976, ii–iii)

Despite the wide use of the term there is really no such thing as 'automated cartography'. It is true that many mapping agencies are automating processes previously carried out manually, some of these using computer technology; but the computer will never totally replace the cartographer. The computer will, however, relieve the cartographer of many onerous repetitive tasks and give a great deal more time for, and assistance to, creativity. Automated cartography suggests a replacement of man by machine; computer-assisted cartography suggests the creative use of the machine by the cartographer, which is both philosophically and methodologically more desirable.

Computer-assisted cartography is a general term, and one might question the utility of such a broad definition and seek further refinement. There are a number of directions such refinement might take, such as the varying degree to which the computer is used or the ways in which the computer is used to help produce maps.

It is possible to distinguish, in actual map production today, between two very different types of computer-assisted cartography: automated mapping and computer mapping. Automated mapping is defined as the automation of map-making processes with an aim to produce maps not inherently different in style, design and content from existing topographic maps. Computer mapping, on the other hand, is defined as the production of maps utilizing primarily the analytical power of the computer. These usually tend to be of a thematic, socio-economic type which rarely reach the high standard of precision of those produced by automated mapping and which may, in fact, be a very different product from the conventional map. Although two types of computer-assisted cartography are distinguished, these should be viewed as being at either end of a continuum. There are systems which combine elements of both and it is theoretically possible to add a computer mapping element to automated mapping and vice-versa. There are examples of such combinations but research and production in computer-assisted cartography today tends to lean towards one end of the spectrum or the other.

This, in many ways, must change if significant progress is to be made. Morrison (Chapter 2) sees the present situation simply as a stage in the development of the discipline; Rhind (Chapter 3) considers such a division unhelpful; Harris (Chapter 5) argues that topographical mapping must provide a digital base to avoid a situation which he sees as wasteful and confusing where thematic mappers make their own digital base map; whereas Bie (Chapter 7) poses the problem in terms of whether the soil surveyor is primarily a boundary delineator or a collector of basic data. Digital data bases today abound, and will probably proliferate at an exponential rate. Bickmore (Chapter 11) identifies research on the cartographic data base as critical to the future health of computer-assisted cartography. The computer provides an opportunity to inter-relate different data bases but that opportunity may disappear if various agencies produce and structure data bases to meet their own requirements without reference to longer-term needs and the requirements of others, which is unfortunately too often the case. The problem here is not primarily a technical one, although there are clearly differences in data structuring between locational and numerical data. The heart of the problem is administrative, philosophical and political. The computer did not create the need for cooperation among cartographers and cartographic agencies: it simply highlighted the existing problems and made their solutions more pressing.

This is, however, only one of the many issues raised by this book which draws together contributions from an international group which has been actively involved in research and development in computer-assisted cartography for many years.

Morrison and Rhind consider, from very different perspectives, the general nature of computer-assisted cartography and its impact on the discipline. Boyle looks at developments in equipment and techniques indicating that there have been major changes since 1976. Technology in the computer field changes at incredible speed and it is difficult to keep pace. Indubitably, some of the material in this book will be outdated between the time of writing and the time of publication. Harris discusses developments in topographic mapping by computer which is complemented by later chapters on soil mapping and geological mapping by Bie and Gold respectively. These illustrate the strong linkages between the environmental sciences and cartography emphasized by Bickmore. The chapter by Ottoson and Rystedt is especially interesting as it indicates how in Sweden, one of the pioneers of computer-assisted cartography, a greater degree of integration has been achieved between automated and computer mapping than elsewhere. It also illustrates the advantages of a step-by-step approach to the introduction of the computer and the advantages of a mix of manual and computer methods. Broome and Witiuk describe the approaches used in mapping census data by computer, and emphasize the difficulty in merging statistical and topographic data bases. Waugh takes us through the development, implementation and application of a typical thematic mapping system, and the volume concludes with Bickmore's thoughts on future research and development in computer-assisted cartography.

Each author has taken a critical look at an area in which he has special expertise. To facilitate comparison, the chapters have been structured around the simplest possible common framework of computer-assisted cartography: Data Capture, Spatial Referencing, Data Manipulation, and Data Display. Where technical terms have been used an explanation has been included although there is as yet no general agreement on common terminology. This, in some senses, is the result of the relatively recent advent of computer-assisted cartography and the dynamic ongoing development in the field. Each author was also asked to specifically address the topic of costs and although this is a contentious area on which it is difficult to reach agreement, some recent estimates have been produced. Opinions on this topic in the book vary but the bulk of the evidence supports the view that the cost *impact* of computer-assisted cartography is a positive one.

The overall impact of the computer on cartography has both positive and negative aspects and throughout the book both of these are considered. Differing opinions appear, as is only to be expected on contentious issues. The nature and definition of cartography itself is a contentious issue and cartography is in the midst of a revolution of which computer-assisted cartography is an important part.

Differing views on the nature and purpose of cartography appear in this book; sometimes explicitly, but more often implicitly. One of the more comprehensive definitions of the discipline is that of the International Cartographic Association which defines cartography as, 'The art, science and technology of making maps together with their study as scientific documents and works of art. In this context maps may be regarded as including all types of maps, plans, charts and sections, three-dimensional models and globes representing the earth or any heavenly body at any scale' (I.C.A., 1972, p. 2). Within this broad definition, however, there are many possible interpretations, the most fundamental of which is whether cartography is a discipline or merely a set of techniques. A growing body of opinion would agree with John Wolter that cartography is in fact an 'emerging discipline' (Wolter, 1975) and that the computer, as Morrison argues, is having a fundamental impact on the form and speed of that emergence.

Along with the differences of opinion in this book there are also agreements. There is a consensus on the vital role of education in modern cartography. Modern cartographers must be knowledgeable about the computer and its impact, and there is a need for education in the widest sense of the word in computer-assisted cartography. It is to this process that this volume will hopefully make its most significant contribution.

This introduction was written after the chapters were all received by the editor and its brevity is indicative of the comprehensive coverage given by the contributors to the main purpose of the book. There is little need to reconsider material which will be more than adequately dealt with in subsequent chapters.

REFERENCES

International Cartographic Association (1972). *Multilingual Dictionary of Technical Terms in Cartography*. Wiesbaden, Germany.

International Cartographic Association (1976). *Glossary of Automation Terms in Cartography*, second edition (English). Menlo Park, California.

Wolter, John A. (1975). 'Cartography—an emerging discipline', *The Canadian Cartographer*, **12**(2), 210–216.

The Computer in Contemporary Cartography
Edited by D. R. F. Taylor
© 1980 John Wiley & Sons Ltd

Chapter 2

Computer Technology and Cartographic Change

JOEL L. MORRISON

We have read repeatedly that computer technology, both hardware and software, has had a profound effect on cartography. What is usually stated following such a general statement is an example outlining the specifics of the computer's effects on one relatively small procedure or routine in cartography; in short, one specific change is described. This chapter seeks to survey the effects of computer technology on the entire cartographic discipline and tries to characterize the current impact of the discipline's use of computer technology.

WHAT IS THE COMPUTER TO CARTOGRAPHY?

Computer-asssistance represents a completely new and different technology for the cartographic discipline. It comes at a time when the discipline has barely accommodated all the changes in its existing technology that was 'new' after World War II. Along with the myriad of smaller innovations appearing almost continuously since World War II, technological 'change' has been almost 'constant' in cartography. But the advent of computer-assistance has touched each and every fundamental part of cartography and as such it represents truly a revolutionary technological change for the discipline. The past 30 years has witnessed rapid technological change in cartography, an old and tradition-bound discipline. Today, rapid 'change' has become an accepted fact of life within cartography.

Beginning by 1960, and spurred on by the Vietnam debâcle and space exploration, this second and entirely new computer-assisted technology has become available to cartographers. As .with most drastic technological change, its implementation in cartographic practice can be characterized by three stages: (1) reluctance to use, or fear of the unknown; (2) use of the technology to produce replicates of products produced by the previous technology using previous methodologies; and (3) full implementation including expanded potentials for the discipline. Every cartographer and cartographic unit must pass through these stages, and of course all do not traverse the same stage coincidentally or at the same speed. Therefore, if searched for, examples of cartographers or cartographic units in each stage could probably be found in the U.S.A. and Canada today. Overall the centre of gravity of the discipline is probably nearing the end of Stage II and is ready to embark on Stage III. However, to begin this survey of the effects of computer-assisted technology on cartography let us first summarize each stage.

Stage I (early 1960s)

Rapid developments occurred in the decade of the 1960s in the writing of computer cartographic algorithms. Prior to 1960, most work of use to cartographers was done by computer manufacturers. During the 1960s, with the aid of computer manufacturers, government Research and Development units moved into cartographic computer-assisted algorithm specification. Certain individuals in cartographic research units at universities also began using the computer-assisted technology at this time. For the most part, however, university teaching, government production units and especially private cartographic firms were unaffected by the developments during Stage I. Many adopted a 'wait-and-see' attitude.

Stage II (late 1960s and 1970s)

As the discipline neared the end of the decade of the 1960s and entered the decade of the 1970s, the centre of gravity of the discipline moved into Stage II. During this period algorithms developed by computer manufacturers or by government or university research units, were traded, bought, copied and translated freely for use on various hardware configurations. Acceptance of computer-assistance in cartography to replicate products done previously by hand became widespread. Initial attempts to replicate existing cartographic methodologies by machine which did not result in quality products were not readily accepted by established cartographers. Many of these initial attempts *were* readily accepted by non-cartographers who had never before utilized maps as fully, due either to lack of knowledge, inability to operate within the time span required for manual map production, or cost factors. The move to the use of computer-assisted technology in Stage II ultimately served the very useful purpose of expanding the community of users of cartographic products.

As Stage II developed, however, replication with quality equal to or exceeding that possible with manual production became a real possibility. Thus most established cartographers, who once were reluctant to consider computer-assistance, began to accept computer replication of previously done manual map-making routines.

Stage III (post-1980)

Stage III is still on the horizon for most cartographers. A few venturesome souls have speculated about the full use of Stage III but the discipline as a whole is basking in the new technology's ability to replicate, at lower costs, previous manual mapping methods. A few new products have been introduced during Stage II and they are finding acceptance; for example the orthophotomap and the temporary map. Stage III will result in more new products that will be more readily accepted because the path to acceptance has been previously demonstrated to be worthwhile.

The purpose of this chapter is not to speculate on the full implementation of Stage III but rather to look at the effects wrought on the discipline by the adaptation of computer-assistance during Stage II. This adaptation has made Stage III a possibility. After a brief discussion of the factors causing changes and the resulting effects, a detailed look at the methodological changes themselves will be taken. The final section will assess trends that are evident today.

WHAT FACTORS ARE CAUSING CHANGES?

Speed of New Technology

Anyone who first utilizes the new technology may take issue with the demonstrated fact that computer-assisted technology in cartography has resulted in speeding-up the process. In fact, the first map produced by an installation, or the first map produced by newly written software, is usually not speedily accomplished. However, after the 'bugs' are out of the system there is no question that computer-assisted cartography is faster than comparable manually produced cartographic products. Draftspersons simply can not match computer-driven plotters in the speed of rendering. Today most standard computer-assisted routines are widely available and only those cartographic units (mainly private firms) that failed to put forth the required initial capital investments are not reaping the benefits of increased speed.

Accuracy of New Technology

Cartographic products produced by computer-assisted technology can usually be made accurate to the resolution of the machine hardware used to produce them. Manual production can never be assured of similar consistency of accuracy. Resolutions are small in modern computer hardware systems as Table 2.1 details. [On the left of Table 2.1 is a series of scales (R.F.); across the top a series of available hardware output resolutions.] It can be seen that with the exception of line-printer maps, resolutions of computer-assisted hardware are capable of matching or exceeding accuracies of manual drafting. Specifically view the coarsest resolution drum plotter, the 0.01 in drum plotter. Even at a scale of 1/100,000,000 this resolution represents a mere 25.4 km on the ground. At our current topographic scales, 1/25,000, the resolution for the coarsest drum plotter is 6.35 m or 20.83 ft on the ground.

When we examine the finer resolution flatbed plotters, topographic scale (1/25,000) resolution is 31.75 cm or just a little over 1 ft on the ground. Resolution, therefore, for the most part, exceeds our data accuracy. Hence computer hardware resolution is greater than that required for most mapping situations given the

Table 2.1 Ground distances corresponding to various resolutions at various scales

R.F.	Typical resolutions	Line printers			Common plotters			
		1/6 in. (4.233 mm)	1/8 in. (3.175 mm)	0.1 in. (2.54 mm)	0.01 in. (0.254 mm)	0.005 in. (0.127 mm)	0.001 in. (0.0254 mm)	0.0005 in. (0.0127 mm)
1/1000		4.233 m	3.175 m	2.54 m	25.4 cm	12.7 cm	2.54 cm	1.27 cm
1/2500		10.5825 m	7.9375 m	6.35 m	63.5 cm	31.75 cm	6.35 cm	3.175 cm
1/5000		21.165 m	15.875 m	12.7 m	127 cm	63.5 cm	12.70 cm	6.35 cm
1/10,000		42.33 m	31.75 m	25.4 m	2.54 m	1.27 m	25.4 cm	12.7 cm
1/25,000		105.825 m	79.375 m	63.4 m	6.35 m	3.175 m	63.5 cm	31.75 cm
1/50,000		211.65 m	158.75 m	127 m	12.7 m	6.35 m	1.27 m	63.5 cm
1/100,000		423.30 m	317.5 m	25.4 m	25.4 m	12.7 m	2.54 m	1.27 m
1/250,000		1058.25 m	793.75 m	635 m	63.5 m	31.75 m	6.35 m	3.175 m
1/1,000,000		4233 m	3175 m	2540 m	254 m	127 m	25.4 m	12.7 m
1/5,000,000		21,165 m	15,875 m	12,700 m	1270 m	635 m	127 m	63.5 m
1/10,000,000		42,330 m	31,750 m	25,400 m	2540 m	1270 m	254 m	127 m
1/50,000,000		211,650 m	158,750 m	127,000 m	12,700 m	6350 m	1270 m	635 m
1/100,000,000		423,300 m	317,500 m	254,000 m	25,400 m	12,700 m	2540 m	1270 m

currently available data. Finally, compared to the perceptual limitations of most map-readers at any given scale, these potential resolutions of computer hardware exceed most observer's abilities to scrutinize the data. The current need, therefore, is for no further improvements in machine resolution for normal cartographic output purposes.

Cost of New Technology

True to most new technologies, computer-assisted cartography does not become cost-effective until late in the developmental game. Once it is in a full capacity production setting, it can be tremendously cost-effective. Two factors mask this cost-effectiveness, however: (1) the initial capital costs and the development of production-oriented software cause initial cost ineffectiveness; and (2) the capacities of the hardware are usually so great that a production unit rarely reaches 100 per cent production capability early in its employment of the hardware. Overall, the computer-assisted technology has reduced the unit costs of comparable manually produced maps.

Increased Output possible with New Technology

Another factor must be mentioned; that is, computer-assisted technology presents the cartographer with a greater array of possible choices. It increases the availability of maps for the masses. Just as we have indicated above that the new technology permits cheaper maps in less time, it stands to reason that more maps for a fixed total cost can be produced in the same time. This increases the cartographer's flexibility in map-making. He can literally make several maps for the same cost and time required for a manual map and choose the one he likes best. In one sense it may cause map proliferation just like fast paper-copiers have caused paper proliferation.

Another Factor of Note

An additional factor to consider is that computer-assisted cartographic technology has developed and has been implemented fastest in government agencies. Whereas in manual cartographic production, private firms may have been the most efficient in terms of costs and time, most private firms have not yet moved into computer-assisted environments. The manual units of private firms are still cost-effective for the most part. The governmental mapping units, on the other hand, who do not have the demands of efficiency or are not required to make a profit, have experimented with the new technology and are thus now leading the cartographic discipline in computer-assisted production capabilities. Undoubtedly private firms will soon move into the use of computer-assisted technology, and spin-offs from the governmental sector experience will make the private firm's utilization of computer-assisted technology even more efficient.

In summary, computer-assistance has speeded cartographic production, increased its accuracy, lowered its unit costs, and increased its flexibility; not a bad achievement. But what has been the effect of these achievements on cartographers and their discipline? The next section will attempt to outline the major effects that have resulted from the switch to computer-assisted technology by cartographers.

WHAT EFFECTS HAS COMPUTER-ASSISTED TECHNOLOGY HAD?

Emergence of Self-identification

The change in technology to the use of computers has speeded a trend that was already beginning to manifest itself within the cartographic discipline. That trend is the one towards the identification of a separate and distinct discipline of cartography. Wolter (1975), in *The Emerging Discipline of Cartography* demonstrates that cartography has all the prerequisites of a separate and distinct discipline. He cites the dramatic rise in numbers of professional societies of cartographers, a similar rise in the number of cartographic periodicals and the great expansion in the volume of cartographic literature that has taken place since World War II. Computer-assisted technology within cartography has not been the source of this emergence, but certainly the new technology has speeded its emergence. An increase in the number of professionals interested in cartography and the growth of training programmes are two further indices of the emergence. Computer technology has without a doubt stimulated both of these areas.

Precise Definitions

Exactly how has the new technology speeded this emergence? Perhaps most importantly it has required the cartographer to define terms to a degree of precision never before considered necessary. Similarly, cartographic methodology has had to be reduced to a finite number of steps which are precisely defined. By meeting these requirements, the cartographer in effect was forced to define in detail all definitions and methods of procedure. Once the methods were completely outlined, more precise definitions of terms were relatively easy to produce. However, acceptance of the more precise definitions leads to another problem as some professionals continue to use terminology in a former, less strict, sense which is creating confusion amongst the members of the discipline. Take the often used words of generalization, classification, or interpolation: cartographers manually performing these processes often never bother to define them precisely. The processes are subjectively carried out, and replications of the results are impossible. In computer-assisted cartography this method of procedure is unacceptable. Terms must be precisely defined for cartography performed with computer-assistance to be replicable.

This need for precise definition has led to further speculation on a more rigorous definition of cartography itself, which has been beneficial to the discipline. Still, some cartographers persist in decrying this trend. The fact remains that a map produced with computer-assistance has the same authenticity that a manually produced map has; however, the former can be replicated and its methodology is more precisely defined.

Freedom from Tedious Production Tasks

A second important way in which computer-assisted technology has affected change has been through freeing the cartographer from the drafting table and allowing the accruing free time to be spent thinking about 'how' and 'why' the map is being designed. The overall scheme of activity thus becomes more readily apparent to the cartographer. Rather than doing, a cartographer can adopt the questioning stance of 'why' or 'what' am I doing? At the same time computer-assisted cartography gives the cartographer more options. Thus using computer-assisted technology the requirement for a clearly planned display resulting from the cartographer's efforts becomes paramount. The cartographer has the necessary options available and the time to produce a first-class product that will communicate his intended message using computer-assistance.

New Products

Computer-assistance has also given the cartographer new products with which to work. Temporary maps that occur but briefly on a screen, and video displays that can be rotated, dissected or manipulated in many ways, along with synchronized real-time displays, all can be added to the cartographer's collection of methods which can be used to convey a message.

Thus at a time when the cartographic discipline is emerging, its emergence has been hastened by a new technology that requires precise definition of all terms and methods, frees the draftsperson from monotonous, time-consuming hand labour, and offers additional options and methods for use. In all likelihood cartography would have continued its growth towards self-identification, but the slope of the growth curve has certainly been steepened by computer technology.

Negative Effects

All of this recent rapid change has not been without its drawbacks. Confusion as to the exact nature of a map has arisen. Such questions are posed as: Is a photo a map?; Do topological data structures constitute an essential cartographic subject for training? Practical answers abound but the theoretical questions remain as yet unanswered. While these questions may be disturbing to some, they are essential to the further development of the cartographic discipline. If computer technology has forced cartographers to answer these questions, then so much to the credit of the technology since cartographers will ultimately stand to gain much by agreeing to precise answers to these fundamental questions. The philosophy of cartography needs postulation. One current perception of cartography is that of a vibrant, exciting field experiencing radical change. Most professionals see cartography as such. Others, outside the discipline, unfortunately sometimes still see cartography simply as a tool or technology useful to further their own discipline. Many disciplines still utilize cartographic methodology without knowing it. In characterizing the current identity of the discipline, Wolters' use of the term 'emerging' is entirely appropriate. Computer technology has hastened this emergence.

WHAT METHODOLOGICAL CHANGES HAVE RESULTED?

While computer-assistance has hastened the emergence of cartography it is also fostering methodological changes. One convenient way to discuss these methodological changes is to categorize them under the headings of data gathering and storage, compilation, generalization, lettering, and production. In each of these areas specific methodological changes can be traced to the use of computer assistance.

Data Gathering

When computer hardware was first available for cartographic use, the need was for data. Data simply did not exist either in sufficient quantity or quality for the cartographer to utilize fully the capability of the hardware. Manual cartographic techniques utilized either other maps, censuses, or air photographs as data sources. These sources were relatively unstandardized in format, scale, extent and time of coverage, and quality. Suddenly machines were available which could utilize larger quantities of data but required that the data had to be in a pre-specified form. Two principle data types were developed for use in computer-assisted cartography: (1) raster data; and (2) stream data.

Raster data consists of a detailed matrix of numbers referring to small geometric areas. Row upon row and column next to column of these data are joined to create a cartographic image. Location is referenced by row/column positions in the matrix and in the simplest case the matrix is binary, consisting of 0 or 1 denoting the absence or presence of some phenomenon. The matrix represents the distribution in machine-useable form.

Stream data, on the other hand, consists of a string or chain of $x-y$ locational coordinates. The trace or path obtained by connecting successive coordinates with straight lines will reveal the mapped phenomenon. Areas are denoted by their boundary lines and points are referenced by a single pair of $x-y$ coordinates.

With the advent of these two types of computer-compatible data gathering, the cartographer has found himself with a new set of problems. Data capture, which had been a rather subjective art when maps were produced manually, has become a major cause for concern in the computer-assisted age. How the data should be captured and further how it should be stored (see below) have become primary concerns. Digitizing has become frequently employed and many tedious hours of work, manually tracing features from maps, are required to convert existing mapped data into machine-usable form. The cartographer freed from the drafting table by computer-assistance is suddenly chained to the digitizing table although recent technological advances as described by Boyle in Chapter 4 may change this. Literally the cartographic world has set out to digitize the real world with all its complexities.

The methodological changes brought about by this need to have all data in one of two forms only began with the digitizing of data. Quickly the problems of conflicting data sources reappeared, only this time in the form of unreconcilable digitized data. Secondly the incompatibility of raster and stream digitizing created further data problems. Simply stated, computer-assistance created a revolution in the methodology of data capture. The manual techniques of data compilation by projection onto base maps, or by tedious scale changing, by pantographs, etc. were no longer problems: instead the incompatibility of raster data and stream data was the major problem. Determination of the scale of input for stream digitizing and of pixel size for raster data captured the attention of cartographers. The cartographer, carefully manually repositioning a road and railroad next to the coastline so as to prevent the lines from touching or overlapping on the final map, gave way to precisely prescribed methods for dealing with such problems in the abstract and at the stage of data gathering and storage, since a cartographer was no longer able to subjectively interfere in the machine drafting production process.

Today a new problem in data capture is evident. We have the capacity to collect data that far outstrips our needs for it. We are in danger of being inundated by machine-compatible data. If we are not crushed by the sheer bulk of machine-usable data at least it is serving to destroy the efficiencies in time of computer-assistance. Time is wasted, costs are increased, and flexibility is forfeited as a result. Common sense and a judicious look at actual user requirements are the preventive medicine for this problem.

Data Storage

Whereas a map library was the excellent place to store cartographic data prior to computer-assistance, digitized map data has different storage requirements. Cards, tapes (paper and magnetic), disks, chips, etc. are all data storage methods for digitized cartographic data. Furthermore, just as some cataloguing or classification system increases the use-potential of a map library, so it is necessary to have indexes of cartographic data stored in machine-useable form. The storage system for data in computerized form is called a data structure. On printed maps, the locational reference in its conventional form serves as the organizational mechanism for storage. In computerized data, the data structures perform that function. Hence the cartographer has another different set of problems with which to deal.

When cartographers were still in need of data in computer-compatible form, no one paid much attention to the way data were digitized and stored. As more data became available the incompatibility of various storage methods often required the complete restructuring of one data set to allow its use with another. This was inefficient, and defeated the very advantages of computer-assistance. As data bases became huge these inefficiencies could no longer be tolerated and precise data structures to ensure efficient operation of the system were needed. The entire topic of cartographic data structures is currently receiving much attention and represents a new area of concern in cartography. The sheer magnitude of data available today makes this necessary.

Literally cartography is on the verge of being inundated with machine-readable data. If it is not stored in a manner that can be accessed efficiently some of our gains brought about by computer-assistance will be lost. As with data capture, our technological capabilities to store data probably are outpacing our needs for these data in storage. We are saving too much data. Digitizers can currently capture data variations at the scale of digitization that exceed our perceptual limits. Why save these data? Calculations and measurements resulting from such data are no more precise than the same calculations and measurements taken from half the amount of data in some cases, and in almost every case the precision of the lesser data set exceeds what we need for answers to our questions. Not only efficient data structures but also sane amounts of data are called for in modern cartography.

Also necessary is the ability to move quickly and efficiently between raster and stream data. Structures for raster data are more reminiscent of map format structures; the same is not true of stream data structures. The methodological changes in data storage then are simply: (1) an increased importance to the structure of the stored data; and (2) the saving of a sufficient but not excessive amount of data. No cartographer utilizing computer-assistance can safely ignore data structures or the amount of stored data today.

A final methodological change in the area of data gathering and storage is revealed by the change in editing procedures. In manually produced maps editing is essentially ideographic, that is as each map is produced it is edited. In computer-assisted cartography, data entered into machine storage are edited without respect to the final use; that is, the material is nomethetically edited.

Computer-assisted cartography has placed greater emphasis on data gathering and storage procedures. New problems must be overcome by cartographers, the procedures must be repeatable and not subjective, and the results must be universally useable; not uniquely so.

Compilation

Two major methodological changes have occurred in compilation. First, the sometimes tedious manual procedure of compilation has been replaced by a rapid, rather simple, automated one; and second the cartographer now enjoys much more flexibility in compilation procedures than dreamed possible when only manual methods were available. This increased flexibility requires the cartographer to think about 'what' he is doing rather than merely to do it. The actual execution is carried out almost instantaneously with computer-assisted technology.

Standard manual compilation techniques required the cartographer to select information from a variety of sources, often at different scales and on different projections, and to compile this information into one common map framework. Several pieces of equipment could be used to aid him in performing this procedure. These include: a Saltzman projector, a pantograph or a Kelsh plotter and others. Enlargement or reduction could be accomplished by other methods as well, but the end result was usually a unique worksheet which could not be exactly replicated if lost or damaged. The procedure was subjective and much of this subjectivity resulted from the level of knowledge the compiler had about the distribution being compiled. Essentially the compilation process was dependent on the subject knowledge and the drafting skill of the cartographer.

The first step in the manual compilation was often the plotting of a projection. When manually compiled, one of the major concerns in the projection selection was the ease of its construction. For example, a rectangular equal area projection was far easier to construct than was an azimuthal equal area. Not surprisingly the former was used more often than the latter. Manual compilation thus tended to keep to a minimum the number of projections that could be used from a practical point of view. Automated compilation, on the other hand, does not distinguish between projections on the basis of ease of construction. The computer-driven plotter can plot the graticule of any specified projection faster and more exactly (precisely) than any projection can be plotted manually. The difference in the time it takes computer-assisted equipment to plot one graticule over another is not significant. Thus the cartographer is no longer constrained in projectional system selection by a concern for the complexity of projection construction. Instead a cartographer needs to know how to select or define a projection which enhances his map's purpose. This new flexibility offered by automation thus gives the cartographer more systems, which in turn requires the cartographer to scrutinize more closely the properties of the various systems of projection. The compilation of a coordinate system is similarily facilitated by computer-assistance. The additional flexibility extends to the fact that everything is repeatable should it become damaged or lost. Furthermore, if the first selection is not to the cartographer's liking, a new selection can easily be made without a tremendous investment in time and money.

Both base and thematic data are also easier to compile using automation. If the data have been accurately digitized and stored, each data set can, with computer assistance, be compiled on a map base in a matter of minutes, if not seconds. An entire worksheet may never actually be compiled in automated production since, especially if colour is being used for the final map, each separation can be compiled directly onto a medium used in photographic processing and/or printing.

Thus the once-tedious chore of compiling a worksheet, consisting of constructing a projection and the compilation of data from different sources and scales, has been changed to a rather rapid step in the map-making process. In this step using computer-assistance a repeatable product is created which can be easily changed, if need be, prior to finished map production. The cartographer therefore shifts his concern in carrying out compilation by using computer-assistance to one of selecting from amongst many options, with the possibility of changing the selection rather easily and rather late in the map-making process (for example, a cartographer may make two or more compilations in radically different formats to see which is preferred). The cartographer with computer-assistance is no longer 'locked in' to one compilation/projection scheme from the beginning of his compilation stage.

Generalization

The detailed methodological changes resulting from automation under the heading of generalization fall principally into two categories: (1) statistical manipulation of the data; and (2) symbolization. The statistical manipulation of data includes both of the generalization elements of simplification and classification. Anyone who has manually prepared a dot map from census data will readily appreciate the many benefits of automation in this area.

Using manual production techniques it is difficult to distinguish between simplification and classification processes. Usually a line or boundary is the manual product of a combination of both process types. This intermixing of the processes, along with the inherent biases added by the incomplete and uneven knowledge of the cartographer, make it almost impossible to replicate a manually derived line or boundary generalization. With computer-assistance each generalization process has had to be defined and the result has been a clearer distinction between the simplification and classification processes, and repeatability of the results. What has been lost is the detailed knowledge which an expert on a given distribution can bring to the manual application of the processes. Attempts at simulations of this have been made. Manual override mechanisms of machine processes are possible on some systems, and by using override capabilities an expert on the data being mapped can monitor the processing and alter the output where the algorithm appears to be producing spurious results. Even with override features the time factor between start and completion has been radically reduced for the entire process by use of computer-assistance.

A few examples will serve to point out the nature of these changes. First the time needed to compute a mean and standard deviation for a group of 3075 observations has been reduced to seconds. The classifying of each observation into a category based on these calculated parameters, or others, can follow as quickly. The resulting map pattern can be plotted and viewed, and if unacceptable, a new class interval scheme computed and displayed all within the time it takes to compute the mean manually. Further the introduction of sophisticated statistical classification schemes is also possible with computer technology. Prior to computer-assistance one could not map the results of a factor analysis very easily because the number of calculations involved was prohibitive. The same was even true of residuals from regression. Autocorrelation functions for surface distributions could not be calculated to aid in the selection of interpolation models for isarithmic line production. In general, computer-assistance has given the cartographer more data-manipulation options and at the same time has speeded their calculation. The result is that the cartographer now must know these statistical methods in order to make an intelligent selection of which one to use in a given map-making situation.

Similar to the classification processes noted above, the simplification processes performed with computer-assistance enable the cartographer to attain a consistency of generalization that was impossible before. This consistency can be accomplished quickly and is repeatable; it is useful not only in map preparation but in the editing of digital files that will be held in storage for future use. This is an important point and a further change in the methodology used with computer-assistance. Simplification processes are now necessarily and routinely applied to data prior to entry into machine storage to reduce the vast amount of redundant data captured in the data-gathering stage. This of course has no direct counterpart in manual production methods.

For data-manipulation schemes, therefore, computer-assistance has: (1) increased the number of options available to the cartographer; (2) resulted in a degree of consistency of treatment not possible in manual cartographic production; (3) accomplished these tasks in less time than they can be accomplished by hand; and (4) allowed for the precise replication of results and quick re-design or evaluation if necessary. These changes require the cartographer to know less about the specific details of the phenomenon being mapped, but more about the appropriateness and the application of the statistical methods employed on the data. These changes serve the cartographer by postponing the 'lock in' point from the beginning of the map preparation to a much later state in a given process. For example if, after viewing the plotted results, a cartographer does not like the display, it can easily and quickly be re-done utilizing variations in technique. This would be impossible to do manually without a great waste of time and energy.

The other major category of change, symbolization, has been less affected by computer-assistance than have many of the other categories discussed above. It is safe to say that any symbol produced by hand can also be produced with computer-assistance. Additionally, computer-assistance can ensure that a symbol is consistently drawn. The speed with which a data set can be symbolized is greatly increased with computer-assistance, especially if the symbol is intricate. The methodology, however, has not been drastically changed.

Point symbolization by machine suffers from placement problems. A visually checked, manual override on the computer allows the cartographer to place the symbols appropriately but slows down the process. Completely automatic placement of point symbols, especially in areas of great symbol density, has not been perfected, however. A rather easily operated manual-override check is usually necessary. For line symbolization or area symbolization by computer-assistance the quality generally equals or exceeds the corresponding manual symbolization. Registration problems can be handled easily by equipment driven by computers. The placement problems encountered in point symbol placement are usually not present in line and area symbolization.

Therefore the symbolization processes can be done by machine production but the savings are not as great as in other areas. The cartographer is freed from the manual production labour that was required of manual symbolization, and this time can be spent ensuring that the symbolization being used is theoretically correct and perceptually accurate. Cartographers, therefore, must still know how to appropriately symbolize data measured on various scales of measurement. Some of the savings in time by use of computer-assistance are lost due to the necessity of visually checking and manually overriding point symbol placement, however.

Lettering

Lettering suffers from the same problems as point symbolization, only the problem is more critical. Any manually produced letter, as with point symbols in general, can also be machine-produced, but letter placement remains the problem. Manual intervention in the machine placement of lettering is therefore still a necessary task. This problem is not inconsequential. It has been estimated that in atlas cartography prior to computer-assistance fully 40 per cent of the manual preparation time is spent with lettering. Because of the importance of lettering and the inability to automate its placement with complete satisfaction, it is not unreasonable to expect that maps without lettering will be commonplace in the future. Temporary maps especially will probably be produced without lettering. Since automatic lettering placement has not been fully accomplished, it is probably true that in current computer-assisted cartographic production, lettering concerns require even greater than 40 per cent of the preparation time.

Computer-assistance can aid in some areas of cartographic lettering, however. For example, lists of all the place-names for a given map area can easily be produced by the machine, as can size variation in letters. These computer aids allow for a greater consistency and completeness in the treatment of lettering in a map area.

In addition to the immediate prospects for less lettering on machine-produced maps, another trend in lettering will probably be to reduce the number of different styles used on any one map. Until rather recently, computer-plotted lettering has looked mechanical and stilted. Although great advances in providing cartographers with more styles and size options have recently been made such as the Hershey founts used in the GIMMS system (described by Waugh in Chapter 10); cartographic design research is indicating that readers can perform as well, if not better, with fewer type styles and sizes on a given map. Hence major changes in the use of lettering on maps may come about because of factors other than just computer-assistance. To date with computer-assistance and due to the placement problem, map-makers and map-users have had to be satisfied with mechanistic-looking lettering, both in style and placement. Methodologically little has changed in map lettering due to computer-assistance. Some major changes may be in the offing, however, and these include trends toward less lettering and the use of fewer styles and size variations on machine-produced maps.

Production

Computer-assistance has allowed the cartographer to expand the range of cartographic products. Virtual maps, coordinates existing in machine storage; temporary maps, maps portrayed for but a short period of time on C.R.T.s; and enhanced images resulting from the machine processing of remotely sensed data; are all examples of new cartographic products resulting from the implementation of computer technology. Each new product adds a new methodology. Some of these closely parallel existing methodologies; others do not. Simultaneous with the creation of these new products and methodologies has been the conversion of the rendering of standard map-production techniques from manual to machine-driven equipment.

The capability to switch from manual to machine-driven production has been adequately demonstrated. Increased accuracy and precision, as well as a reduction in production time, have usually resulted. Of these, the more impressive is the increased precision and accuracy of plotting that has been realized. Manual positioning errors of lines and boundaries have been eradicated. Provided the positional coordinates for the feature are correctly recorded in machine storage, the feature will consistently be accurately placed by computer-driven equipment. The precision of the placement by machine cannot be matched by manual techniques, and the precision of the repeatability of machine placement is unquestionably superior to that of similar manual placement. Computer-driven equipment is available to replicate almost all previous production technology and is described in detail by Boyle in Chapter 4. (Flexibility in lettering placement is the only major exception.) Laser plotters can more quickly and more precisely 'draw' output directly onto film for subsequent printing. Where only a few copies are needed, computer-driven printers (for example Versatek) can produce them without the need for going to plates and a press. Therefore the production aspects of cartography have, to date, taken advantage of computer-assistance to replicate former manual processes while providing increases in accuracy, precision, and speed.

The new cartographic products have expanded the use of cartographic methodology to other disciplines and have decreased society's dependence on the printed map. These new products relieve each map of the necessity to be all things to all users. Cartographic products can now be justified to fulfil very specific needs of the map-user's community. The temporary map, for example, aids the researcher as well as the designer. If a plot of a distribution is necessary, the researcher can have one produced on a C.R.T. screen. There is no longer an obligation to title the display, to include base data, a scale, etc., since it will not be lying around an office for other potential users to see. It will be destroyed after the researcher is finished. Because of the relative inexpensive costs the researcher can have it reappear on the C.R.T. when needed, rather than print it. Such a map can evolve with the researcher's ideas and only the final map, a composite of the researcher's previous attempts, may ever appear in 'hard' copy form with title, legend, scale, etc. This represents a tremendous increase in flexibility for the researcher as well as a drastic methodological change.

Likewise a virtual map can be queried for information for which a map-reader might consult a map. The information can be given exactly as recorded in machine storage, and the reader need not rely on estimation skills from a visual display. Such information as the degree of ground slope at a series of points, elevation above sea level, or undisturbed lines of sight can be answered from virtual map information with no graphic ever being produced. Computer-assisted enhancement of imagery has also increased the repertoire of the cartographer. Imagery can be generalized and distributions delimited by machine. The overlay technique of many land-use planners, urban designers, etc. can be accomplished more speedily than ever before.

In the production aspects of cartography, changes in methodology have principally enabled cartographers to demand greater accuracy and precision plus repeatability. They can perform almost all manual production by machine; going directly from machine-stored data to 'hard' copy output and creating either a useable image or stable film to aid in the printing of the map. The effect of the changing methodology, therefore, is once again to free the cartographer from tedious hand work. A cartographic designer can spend more time planning, designing, and thinking about 'why' and 'what' is being done, not about 'how' it is being done. Additionally, the cartographer has the option of reviewing the results of several alternative designs before deciding on one. This is economically impossible in manual production.

Admittedly as the discipline enters Stage III, many problems remain. What is technologically feasible is not necessarily the current *status-quo* in production agencies. Yet a person not in possession of a day-to-day knowledge of work in large federal mapping agencies cannot help but be surprised at the amount of computer-assistance used in these agencies' mapping activities. Implementation does lag behind feasibility, but to a lesser degree in federal agencies than in private practice. Correctly or incorrectly, U.S. federal agencies have made a strong commitment to digital cartography. A student seeking training in cartography is opting for early obsolescence if he or she does not opt to learn the basics of computer-assisted cartography. The opportunity to learn theoretical design considerations is equally necessary. Both are required, to make a well-trained cartographer. One who specializes in current manual techniques will not be employable by 1990 in the federal mapping agencies.

WHAT ARE THE LONG-TERM EFFECTS OF CHANGING METHODOLOGIES?

Several long-term effects of the changing methodologies resulting from computer-assistance are evident in cartography today. I wish to single out four such effects for consideration in the summary section of this chapter.

Simpler Maps

A trend towards simpler maps is one of the long-term effects of computer-assistance. No longer must every map have a scale, title, a labelled projection, a data source, sufficient base data, etc. This is especially true of temporary maps. Since temporary maps will no longer be printed the user audience can be controlled, and base information that might interfere with the communication of theme data need not appear. In fact the map can be kept incredibly simple, even omitting lettering. This trend is toward special maps to a closely controlled set of users. Eventually this will lead to user-produced maps. Creating your own map, in the colours of your choice, is technologically feasible today. As every user becomes a cartographer (within the constraints that of necessity must be put on such a system) each map will be unique. For example, at the entrance to each national park will be a Dial-a-map booth. The user, for a prescribed fee, can specify which of several features are desired on the map. The data can be kept current; for example, daily road work locations, or sightings of wild animals within the last twenty-four hours, or location of currently available camp sites can be instantaneously presented.

Even on printed maps, computer-assistance will lead to simpler maps if for no other reason than the fact that each map costs less to produce and therefore for a given dollar cost, several specific maps can be made instead of one involved map. Alternatively, more complex manipulations can be performed between data sets, thus integrating the data to a higher level of abstraction for the map-user. The resulting relationships may then be mapped by a simpler method making the communication easier and more direct. Without this manipulation the map-user is left to visually integrate detailed data sets. The message communicated by data sets manipulated by computer-assistance can be made clearer because of their simplicity; thus users are less likely to receive an incorrect analysis or interpretation from the one desired.

Defendable Design

A second major effect resulting from computer-assistance is that every design decision can be based on defendable tenets. With the cartographer freed from map-making tasks, loss of bias will result in favour of those methodologies which are easy to execute. Instead the cartographer can use research results to select the best technique to portray the intended message. Computer-assistance has thus freed the cartographer to use many symbolization options and has also allowed these to be changed easily at a late stage in the map's preparation. Literally, if the map design has a flaw, it can be changed at the last minute. With manual methods the 'lock-in' point for the map's design is much earlier. Last-minute changes are impossible, or nearly so, without a considerable loss of time and effort. With computer-assistance this is no longer true.

Definition and Expansion of Methods and Products

A third long-term effect of the change to computer technology in cartography will be the detailed specification of previous methodologies and the creation of new methodologies. This has started, and can be expected to continue, and as Stage III is fully realized a proliferation of new techniques and methodologies for new products can be expected. First and foremost the methods used by cartographers, some for centuries, have been explicitly defined to enable them to be programmed. This has enabled cartographers to reassess each methodology and to utilize different combinations of them, resulting in new techniques for portraying data. It has led to a deeper understanding of the inherent errors in each method used. It has also given the cartographer numerous additional options in such areas as class limits and intervals, interpolation algorithms, and consistently applied simplification routines.

Additionally new products have resulted such as the virtual map, the temporary map, the enhanced image, and the classless choropleth map. These have broadened the usefulness of cartographic methodologies to disciplines far removed from geography. For example, medical science and chemistry, and physics or engineering all are utilizing cartographic methods on C.R.T.s. A map of the brain or the spinal column is as possible as is the response surface resulting from the combination of two compounds in chemistry. Animated cartography showing the time dimensions is now possible, and practical for perhaps the first time to cartographers. Computer-assistance has therefore served to point out the general nature of cartographic methodology. This trend will continue and hasten the time when cartographic methodology will become accepted as standard *modus operandi* in the science of communication.

Education

Finally the above developments will initiate a trend for educational curricular change. Today's students (tomorrow's cartographers), must be fully prepared to utilize and work in a cartography supported by computer technology. Programming skills, image manipulation, digitizing skills along with a knowledge of data structures, and the available production options made possible by computer-assistance are all necessary aspects of the knowledge of a modern-trained cartographer.

Computer technology is the latest of the revolutions in cartography. Not surprisingly the result is cartographic change. We are still on the threshold of the most impressive changes as the discipline moves into the third stage of use of the computer-assisted technology.

REFERENCES

American Congress on Surveying and Mapping. (1976). *Proceedings of The International Conference on Automation in Cartography, 'Auto-Carto I'*. December 1974, Reston, Virginia. iv + 318 pp. (Automation, Coding, Data, Data Manipulation, Symbolization, Digitization).

American Congress on Surveying and Mapping. (1975). *Proceedings of The International Symposium on Computer-Assisted Cartography, 'Auto-Carto II'*. September 1975, Reston, Virginia. vi + 614 pp. (Automation, Design, Statistical map, Map reading, Perception, Colour, Coding).

American Congress on Surveying and Mapping. (1979). *Proceedings of The International Conference on Computer-Assisted Cartography, 'Auto-Carto III'*. January 1978, San Francisco, California. 520 pp. (Display Design, Data manipulation, Generalization, Coding, Digitization, Automation).

British Cartographic Society, *Automated Cartography* (1974). Papers presented at the Annual Symposium of the British Cartographic Society, Southampton, 1973. Special Publication No. 1, British Cartographic Society, London.

Harvard University Laboratory for Computer Graphics and Spatial Analysis (1977). Papers, *An Advanced Study Symposium on Topological Data Structures for Geographic Information Systems*.

Institute of British Geographers (1977). 'Contemporary Cartography', *Transactions, IBG*, New Series, vol. 2 124 pp.

LeBlanc, A. L. (ed.) (1973). *Computer Cartography in Canada*, Cartographica Monograph No. 9. 103 pp.

Rosenfeld, A., and Kak, V. C. (1976). *Digital Image Processing*. Academic Press, New York, New York.

U.S. Department of Commerce, Bureau of the Census (1978). 'New Developments', *Mapping for Censuses and Surveys*, chap. 14.

Wastesson, O., Rystedt, B., and Taylor, D. R. F. (eds.) (1977). *Computer Cartography in Sweden*, Cartographica Monograph No. 20. 114 pp.

Wolter, J. A. (1975). *'The Emerging Discipline of Cartography'*, Ph.D. dissertation, Department of Geography, University of Minnesota.

Chapter 3
The Nature of Computer-assisted Cartography

DAVID RHIND

INTRODUCTION

In what sense can computer-assisted cartography (C.A.C.) be said to exist? Is the relevant criterion that there are a number of people engaged in making computer systems which produce maps or, alternatively, that some others make some maps this way? Surely this is insufficient; many more people are daily engaged in pouring out cereals for breakfast than make maps using computers—yet we do not grace them with the accolade of being a corporate group or discipline. Is C.A.C. a challenging intellectual focus, currently occupying some of the best minds of our time as, for example, is astronomy? Is it everywhere a factor which is going to change the life of people outside the rather limited numbers of map-makers? Or is it merely a new tool elevated by inflated academic claims to the level of paradigm shift—the cartographic equivalent of all the changes in post-war geography so cynically analysed by Taylor (1976)?

Both positive and negative answers have been given to all the questions above: the situation is therefore a confused one and certainly not one which may be disentangled simply by collecting diverse opinions from the literature, whether these are the latest, the most frequent or simply the most emphatic. What is more, there are clear indications that the acts of individuals or groups working on C.A.C. have some affinity with the physics of a fluid in Brownian motion: if you examine them, changes occur as a result of the examination process. In these circumstances, the theme of this chapter is necessarily one which leads to provisional conclusions and statements of personal belief; appropriately, much of the text is written in the first person. I shall, provisionally and for the convenience of the editor of this publication, accept that some discernibly distinct sub-discipline encompassing C.A.C. does exist, at least at the present time. The essay attempts to establish by precept, by example, and by suggestion, the contemporary and future nature of C.A.C. To set the scene, we begin with five personally held views:

(i) The vast bulk of the research and development work in C.A.C. to date has been of a technical kind: the primary concern has often been to provide technical solutions to limited, practical problems. Few major conceptual advances have been made since the era of Bickmore and Boyle's 1964 paper. Many of the technical improvements which have occurred stem from work done outside the field of C.A.C.

(ii) There are, as yet, few demonstrable and costed advantages of using computers in map-making (eg. Pfrommer, 1975; Thompson, 1978). Most of the so-called advantage studies are based upon comparisons with manual map-making procedures or upon the basis of making many more maps than would have been produced under a manual system; only if one believes that maps are *good things* in themselves or if there is manifest evidence of real need for maps can the latter be seen as a justification. One obvious example of doing things because they are easy rather than because they are wise or necessary is the creation and use of vast numbers of computer-based choropleth mapping packages, even though logic and some cartographic theory (see Williams, 1976) indicates that it is extremely unwise to map many variables in this way.

(iii) Both traditional cartographers and the advocates of C.A.C. have very frequently polarized their views, leading to conflict and, on occasion, difficult personal relationships and consequent managerial problems. Beyond this essentially ephemeral problem, such acts have dichotomized developments in cartography to the detriment of all. Comparatively little interaction has occurred between those concerned with topographic mapping and those with thematic concerns— the difference in levels of supposed adequacy of the end product, the ready availability of line printers in computer centres and the previous tradition of 'apartness' have negated the advantages of collaboration in what is a vertically integrated task—the display of spatial data using some predefined metric.

(iv) Computer-assisted map-making is only one small manifestation of changes in our society which are comparable in scope and performance to those of the Industrial Revolution. These changes are such that we may, in future, reasonably expect a very substantial change in type of work and scope for initiative of the map-maker, in the number and educational attainments of such personnel and in the form of the corporate organizations in which they are based.

(v) Few cartographers have manifested real interest in the social implications of what they are doing: this seems to stem from the traditionalist attitudes advanced by, among others, a former president of the British Cartographic Society. He advocated the viewpoint that 'map-authors' and 'map-makers' were distinct groups. In so doing, he legitimized the craft structure of traditional cartography and maintained cartographers as little more than skilled and thoughtful draughtsmen. To the extent that automation ultimately reduces the amount of manual redrafting to be done, such a viewpoint is a prescription for unemployment. We might note that similar concerns for the future role of a discipline are evident elsewhere: Dale (1978), for example, has summarized the options open to Land Surveyors in the future.

THE EVOLUTION OF C.A.C.

As a basic framework, the following procedures must have been carried out if C.A.C. has occurred: the conversion of geographically identified data into machine-readable form (collected either direct from the field or from such intermediate sources as maps or air photographs); the manipulation of these data, such as changing the map projection, selecting elements and linking data together; and, finally, the plotting of maps, diagrams and, possibly, related statistics. None of these need necessarily have been carried out at the same location, by the same individual, or for the same initial purpose—the widespread university use of topographic data digitized by the Central Intelligence Agency is a case in point.

The first successful attempts to produce graphics from computers were reported in the early 1950s. By the middle of that decade, maps were being produced on the now-standard computer output device, the line printer (for example Dobrin, 1952; Inst. of Met., 1954; Simpson, 1954), on the earliest cathode ray tubes (Döös and Eaton, 1957; Sawyer, 1960) or on tabulating equipment (Perring and Walters, 1962). Both then and now, meteorologists, geologists, geophysicists, plant ecologists, and other earth scientists have been major innovators and users in Britain at least, though a significant growth in use by central and local government planning staff has occurred over the period from 1974 onwards and most national survey organizations in developed countries have at least carried out some experiments with automated mapping. By the end of the 1960s the SYMAP program, created by Howard T. Fisher and developed at the Laboratory for Computer Graphics and Spatial Analysis in Harvard University (Schmidt and Zafft, 1975), was running at more than 100 sites. By 1975 this number had grown to 300; since the majority of these sites were universities, this represented a considerable growth in the availability of automated mapping facilities to academics in general. Universities, however, have contributed comparatively little at the very high-quality end of the spectrum since—though they have often had immense computing power—few have had appropriate output facilities. Since 1975, the evolutionary tree has sprouted so many branches that a detailed overview is not appropriate at present.

C.A.C. IN PRACTICE

The practical capabilities of any C.A.C. system will depend upon its design and upon the amount of human interaction which occurs within it. In principle, however, we might reasonably expect that, from data already stored in digital form, maps could be made of data selected on area or on attribute and plotted with the users' choice of symbolism, scale and projection. We might reasonably expect also that this plotting could be done, subject to the availability of suitable equipment, much faster and at least as accurately as by human means. Finally, we could reasonably expect a well-designed C.A.C. system to ensure that updating or correcting data would be a trivial matter and, in addition, to provide at least rudimentary data manipulation and summary facilities. As Bickmore has pointed out for at least a decade, the essence of C.A.C. is that the data are held quite separately from their representation—unlike traditional cartography—and this provides considerable flexibility in how the data are then used. Sadly, the major practical problem at the time of writing remains the digitizing bottleneck: after running a pilot production line for nearly 5 years, the Ordnance Survey in Britain had 7000 digital map sheets on sale at the end of 1978 yet these constituted just over 3 per cent of the basic scales map coverage of the country.

In practice, about 10,000 people in Britain derive their living primarily from the making or distribution of maps and map-like products: this is of the order of 1 person in 2500 of the working population. Perhaps 500 of these have any on-going connection with computer-assisted mapping and similar (or smaller) ratios probably hold for the rest of Europe. Yet the interest in the topic is extreme: four times as many papers on it were submitted to the International Cartographic Association Conference in Maryland, U.S.A., in 1978 as under any other theme: contributions came from all parts of the world. Meine (pers. comm.) claimed that he held references to at least 3000 articles on automated cartography as long ago as 1975: if this is correct, and on the basis of reasonable trend projections, the total must now be nearing at least 5000. In Britain, of the order of £3 million has been spent by the Ordnance Survey on their digital mapping programme; in very approximate terms, at least £2 million at 1979 prices has been spent by the Natural Environmental Research Council Experimental Cartography Unit on R and D connected with digital mapping: a comparable sum has been spent by the military mapping organization and other organizations, such as Hunting Surveys and Ferranti–Cetec, have invested major sums in developing systems. These conservative estimates, which are certainly dwarfed by some of the North American expenditures, suggest that the field is of more than passing interest to the military and commercial minds and, in particular, to earth scientists.

Yet what has become abundantly clear over the last 5 years is that the use of automation merely as a means of replicating manual processes is not yet always economic: Thompson (1978), for example, has shown that making new 1/1250 and 1/2500 scale maps using the present O.S. digital mapping procedures are, respectively 1.6 and 1.3 times as expensive as their manually produced counterparts, take as long to produce and involve more manual input! Such figures essentially illustrate the state of a map-making process at one particular moment and obvious improvements are already being investigated. In addition, however, such a view is a parochial one: alternative and, occasionally, non-graphical uses of the raw data must be expected although, if new uses, how these are to be costed is rarely clear. In an ideal world, data must be usable for a variety of purposes if the cost of converting it to digital form is to be amortized over numerous heads. For example, the possible uses of digital topographic data can be categorized, as below, mainly in order of increasing complexity:

(i) to plot maps at given scales and for selected areas, showing only selected features and doing this on a selected map projection;
(ii) to provide descriptive statistics, for example of land over 200 m within a drainage basin;
(iii) to recombine functional units, for example build up land parcels;
(iv) to act as a back-cloth, facilitating the input and updating of the users' own data;
(v) to overlay with other data to provide, for example, the number of houses within a smokeless zone or the average slope on one class of land.

The increasing acceptability of this data-related point of view to formerly cartographic institutions can be seen from the statements by the United States Geological Survey (Southard, 1978) and the Ordnance Survey (Smith, 1978) that they are in the business of making available topographic data, rather than necessarily restricting themselves to making available maps derived from these data. Very considerable expenditures are now being made or planned in the creation of digital topographic data bases.

Beyond the provision, maintenance and dissemination of the national digital topographic data base, the mapping of other variables seems likely to become more important: it is already the case that many government-provided statistical data sets in the western world are more conveniently and/or more cheaply purchased in computer rather than paper forms—an obvious example is the U.K. Population Census data. There is an increasing demand for such data for small areas. In the relevant 1981 census data files, some 2600 items of information will be made available for each census enumeration district, as compared with 1571 in 1971. At least eighty-five atlases based on census data have now been published for different parts of the world. In some respects the most impressive productions are those of the U.S. Bureau of Census (Meyer, Broome and Schweitzer, 1975) who produced 5000 colour separations in 14 months. Impressive, for a different reason, are the maps produced by the Census Research Unit in Britain which show up to 150,000 areas on one A4 sheet of paper yet permit many individual areas to be distinguished (Rhind, Visvalingam, Perry and Evans, 1976; C.R.U., 1979). In both these instances far more maps—and probably of far better quality—have been produced than hitherto. Though not as graphically exciting, the results of the Canadian Agricultural Census of 1976 have been produced and disseminated quickly by similar means (Statistics Canada, 1979). Other developments are occurring on a pan-national scale. The European Economic Community, for example, has conducted a series of case studies in each of the national territories on a computer-based environmental assessment scheme, working through the mapping of selected variables, and is moving towards a consistent European Scheme. The eighteen-nation Council of Europe has been discussing the role of automated cartography in regional level planning since *circa* 1973 and is currently (early 1979) formulating plans for implementing mapping of comparable statistics. Given these trends, we may reasonably expect the practical efficacy of a totally computer-based approach to grow in the future.

C.A.C. AND THEORY

'Theory' is a term much misunderstood, especially by cartographers. For example, the vast bulk of entries under the term 'Theoretical Cartography' in the most recent issue of *Bibliographia Cartographica* (Verlag Dokumentation Saur KG, 1978) are classified on the basis that they do not conveniently fall in the other categories; one might doubt whether an article entitled 'Surveying Railway Boundaries' is going to advance cartographic theory a great deal! Characteristically the same, much-used, bibliography includes all explicit mention of automation under 'Cartographic Technology' (even though a typographic error then ensures that much of the English, but not the German, page headings revert to 'Theoretical Cartography'!) Numerous other examples may be given of the widespread identification of automation with non-theoretical concerns.

In saying this, it is as well to appreciate that elsewhere in cartography there has been a considerable growth in theory formulation in recent years. This has included theory on the relationship of cartography to other disciplines and on the nature of perception of maps, both of which have important implications for the way in which cartography is taught, and on what sort of graphic images should be produced. Not all of this has been useful or even appropriate—as is demonstrated by Robinson and Petchennik's (1975) scathing denunciation of the simplistic use of Shannon and Weaver information theory as a communication model for cartography. Indeed, much of the so-called theory concerned with map-perception studies is little more than empirical findings from the application of aggregate statistical procedures, initially designed to deal with readily controlled agricultural experiments. Although there are rare exceptions where such studies do perhaps advance our understanding of 'why', rather than 'how unsuccessfully' people study maps in the way they do, the vast majority of perception studies are of limited consequence to cartography generally and even less to C.A.C. This is because the variance in the responses, even when collected over a short time period and from supposedly homogeneous groups, is often considerable. Unless we can find means of reducing and preferably obviating such variance, such that, say, one method of depiction ensured that the great bulk of the map-users could consistently and demonstrably obtain from the map an accurate representation of what the map creator intentionally and unintentionally intended, cartographic design theory must remain rooted at an extremely broad-brush level and guided, as now, as much by aesthetic as by perceptual considerations. That this contention would not be accepted by all cartographers, however, is evidenced by the furore which arose after Tobler (1973) pointed out that, using automation, there was no longer any technical justification for using discrete classes in a choropleth map. Dobson (1973) and others argued that both theory and perception studies supported the need for such discrete levels of symbolism. Other examples where computer-based cartographers have flouted cartographic convention (sometimes difficult to differentiate from 'cartographic theory') is in the frequent use of proportional squares rather than circles (far fewer vectors need be generated and plotted) and in the use of single bars on point symbols whose length is proportional to the value being mapped (for example I.G.S., 1978).

It seems extremely unlikely to this author that very much more can usefully be achieved by existing style-perception studies: attempts to isolate map components for study at best assume that the interaction effects when all the components are recombined are additive. To me this is highly unlikely and such studies may, because of their simplicity, be biasing our perception of cartographic perception. In so far as C.A.C. can readily provide many different graphic images from the same data, it is thus a potentially dangerous means of facilitating additional, futile empirical studies.

An ultimately more rewarding, if less explored and more academic, area of theory is that concerned with the consideration of maps as a language. This is a long-established notion though, as in Woolridge and East's 'Spirit and Purpose of Geography', was usually stated in non-rigorous terms. Dacey's famous paper on 'The Syntax of a Triangle' and a growing awareness of the modern linguistic theoreticians such as Chomsky have led cartographers such as Youngmann (1978) to attempt definitions of those elements of maps which act as the nouns, verbs, and other linguistic constructs. Less rigorous but more immediately of practical importance is the work on semiotics by Berlin and his co-workers: it is remarkable how no English language translation of *Semiologie Graphique* (Bertin, 1973) has yet appeared, even though the ideas it contains have formed the basis of his small computer-mapping system.

All of the above are elements of theory which might (in some cases ultimately) help the map-maker to understand better what he is doing, or to design maps which convey a particular message more frequently or in a more error-free way to the user. We have neglected, thus far, the role of maps in society, yet to regard them as unambiguous devices for which the cartographer has no responsibility other than the accuracy of placement of line and point features is manifestly mechanistic and short-sighted. There is ample evidence that mapping is a fuzzy means of communicating often-fuzzy data; the propaganda maps, particularly those in pre-war Germany (Ager, 1977) and the many different possible forms of cartographic representation of the same data included in standard texts such as Robinson, Sale, and Morrison (1978)—even though they may contain *ad-hoc* prescriptions on what is to be used for which purpose—testify to this viewpoint. The situation is even more worrying with C.A.C. and, especially, with interpolated data. Rhind (1971) has demonstrated that even the same computer-contouring package can produce very different results from the same data and that considerable understanding of the detailed procedures utilized is necessary to appreciate why some results appear.

Clearly, we have moved away from considering cartography as being merely a draughting exercise based on well-established theory. In these situations it is worth speculating that the kind of maps we use or produce are related to our ideological outlook and to our social and career aspirations, either felt or subliminal. To the extent that we differ in these and our physical characteristics, C.A.C. has much to offer us—personally customized maps can be produced for those males who know they are suffering from red/green colour-blindness and, if it can be programmed, Marxist and Capitalist-style mapping can be produced (probably manifested through the variables mapped?). In so far as cartographic theory is not well established then C.A.C. enables us to produce maps more related to the likes and dislikes of the individual customer. Whether this is desirable or not depends largely upon the view one takes of science (see, for example, Gregory, 1978) and of society, as much as of cartography.

ANTICIPATIONS OF THE FUTURE

If, like Henry Ford, we believe that a knowledge of history is of severely limited relevance to running the present and anticipating the future, it is useful to sketch out, however, tentatively, what the future of C.A.C. might look like. Here some overlap with Bickmore (1979, Chap. 11) is inevitable but this consideration will be restricted to availability and use of C.A.C., rather than to research into it.

The Data Base

We can start with the oft-expressed contention that the conversion of a back-log of existing maps into digital form is a formidable task but essentially a finite one and one which, by the turn of the century, is likely to be well advanced if not complete. In this author's view, the extent to which the digitizing of such intermediate forms of data as the map persists will largely depend upon developments in other, more direct means of data capture and to the stability of mapping units. Let us consider both of these in turn.

Even if a totally new form of data collection is found which meets immediate needs this may still be unusable if heavy investment—either in cash or in human resources—has been put into the old procedures. This is all the more true if the new method does not directly produce results comparable to the old, since time-series data are important for some users. A specific example of this is remote-sensing technology: present 'conventional' civilian satellites and their planned successors—Landsat 1, 2, 3, and D—have a spatial resolution (of about 80 or 30 m) which has virtually no relevance to topographic mapping in Britain: in other parts of the world, in contrast, this resolution is better than anything else which exists. Newer satellite-based sensing, such as the synthetic aperture radar on Seasat, promise better resolution, especially by the mid-1980s, and it is entirely feasible that all the manifestations of change in topography could be detected by such means. However, noticing that something has occurred, and knowing what has replaced what, are two very different things in many instances: field checking, although focused on restricted areas, will still be necessary in topographic data collection unless national survey organizations relax their (often implicit) accuracy of categorization standards. By the same token, much mappable data in the social sciences cannot be surveyed remotely. Remote survey of environmental data is based upon surrogates which seem to vary widely in their temporal and spatial appropriateness. In short, field mapping—and hence a mixed input procedure—may be expected to continue indefinitely although the balance between field and remote survey will change; totally automated mapping systems are a nonsense, especially since considerable human intelligence will be needed for interpretation in the foreseeable future.

In the earth sciences, the boundary is often a major part of the data—one of the two elements ('what' and 'where') which a field surveyor will wish to identify. It is true that these boundaries may sometimes be deduced from data collected on a regular spatial pattern, such as from most remote sensing procedures, but here the boundary positions are much less well identified than in the typically over-sampled manual, field-based procedure. In the social sciences, the boundaries are often given, have no consistent physical expression and merely represent the edges of accounting units—the area within which, for example, one man collects census enumeration forms. Where these areas are stable, little need is experienced for new digitizing; conversely, major changes in administrative frameworks (such as happened in Britain in 1974/75) may require sudden and massive amounts of new digitizing from maps *and* greatly complicate comparisons of change over time in map form. Again, digitizing from maps is likely to be an on-going, if spasmodic, activity.

Mapping from the Data Base

On a personal basis, this author envisages much greater mapping from remote sensing material in the environmental fields and from government-collected data bases in the socio-economic fields than has hitherto been the case—especially in Europe. In isolation, neither of these pose insuperable technical problems. However, if the data are to be combined in any manner other than simple graphical overlay, substantial technical problems arise, stemming from the fact that the only point of cross-reference is their geography, rather than their topology.

Much use has been made recently of the term 'distributed cartography'. In so far as any consistent use of the term has been made, this implies the making of maps on an *ad-hoc*, on-demand, local and personally customized basis rather than on the basis of a central lithographically printed product; occasionally it is also taken to include local collection of data which may be fed into a centralized data base. The technical difficulties of setting this up are rarely appreciated; indeed the difficulties of going one stage further—of having the data base distributed over many computer systems, any of which can access another without having to go through a single central processor, are such that the first announcement of such facilities has only just (March 1979) been made by a major computer manufacturer. Even now, these will not cope with cartographic data bases in which matching of data items may have to be by geographical proximity and error will always be present. It is trivial in most

European countries to be able to transmit simple maps down telephone lines, but we are still a long way from the time when matching of data files and transmission to remote sites can be guaranteed error-free, rapid and cheap.

But let us look ahead to the nature of C.A.C., at a time, perhaps by the early 1990s, when such communication is cheap, largely error-free and plentiful. Provided the televisions of the day have a colour hard-copy unit attached to them, the role of the map printer should be very much reduced. The Open University in Britain already produce a device (CYCLOPS) containing a micro-processor through which it is possible to interactively communicate between users via a standard television set: low-resolution maps may be drawn by light-pen on one tube and seen and modified by the other; information from data bases may also be transmitted to the user, either in tabular or graphic form. Subject to solving the access problems—possibly done through a development of the British Prestel system—and of enhancing the resolution of standard television tubes, the mechanism then exists for producing or even browsing customized maps. Numerous changes must be expected from the present situation—the charging of royalties on browsed maps from which no hard copy is taken, the role of commercial map suppliers in an environment where the state or other monopoly supplier provides the telecommunications links, the protection of confidential data from unauthorized users and the need for tutorial packages to teach yourself cartography via your television tube are only four elements for consideration.

Not all potential map users have a television-set—in Britain this amounts to about 96 per cent of the households; in Portugal, by comparison the figure is about 76 per cent. Nor do all households have telephones. Those figures and the desirability of protecting past investment, rather than providing what is already available in research laboratories, ensures that we are some way off a situation where C.A.C. has moved cartography from being a craft industry to a high technology, specialized and centralized industry to an everyday, every-person practice. At present, this author believes, we are seeing the early stages of an evolution along those lines; one which is both inevitable and desirable. At present, however, we have almost the worst of all worlds—where few users can show concrete advantage from the use of C.A.C., much duplication of work is in progress and where systems have been set up, they exist largely to replicate the usual manual processes. The present distribution of C.A.C. facilities in Britain, for example, is very like that of the first computers which were available to Universities, government departments and major industries in the late 1950s. Now computers are almost ubiquitous and, best of all, few people realize how much mundane work is done by them. If computer cartography can 'ride on the back' of a mass market such as television tubes and telecommunications, there is little doubt that we can broaden its scope and use immensely—at least in some countries. I see no reason why use of a tool of the flexibility and power of C.A.C. should continue to be restricted to elite groups.

Current cartographers may well have misgivings about such a scenario. That is inevitable. It also seems inevitable, however, that a diminution in numbers of lower-grade draughtsmen will be needed in future, though it is totally impossible to quantify the magnitude or timing of this decline. As a consequence, the major mapping institutions may well become smaller and more concerned with data compilation and the monitoring of quality control standards in local offices. In the medium term at least, a growth will occur in the need for people who are capable of using C.A.C. facilities and producing usable maps with them, along with tables of statistics, graphs, and text. This suggestion—and it is really a little more than that—has important, indeed vital, educational implications—the most obvious of which is that all cartographic courses should include some exposure to computer-based methods and (more important) to concepts. All such courses—some of which must of essence be mid-career ones—should encourage the view that change in the tools of trade, in the material to be mapped and in the needs of the end-user will continue. In short, the future adaptability and success of cartography as a discipline is substantially in the hands of those responsible for cartographic education.

CONCLUSIONS

By implication, I have accepted that the activity of various individuals and groups and the spending of money is sufficient reason to accept that C.A.C. may be said to exist as an entity. I have argued that it is (and has been for a decade or more) a corpus of belief, with most individuals passing through and making a technical contribution on their way from and to other disciplines. Traditional cartographers have only rarely made much contribution except as users. Initially, it seems to me, the potential and the use of C.A.C. was a matter of faith, rather than immediate logic. This situation has now changed: the technical aspects are now less exciting simply because many people have worked on them and many technical problems have been 'fixed', if not 'solved' as Boyle indicates in Chapter 4; the systems mostly work. Despite this I have argued that comparatively little operational use is still made of C.A.C. procedures but have suggested that the balance of cost, speed and flexibility is progressively tipping towards the new generations of C.A.C. systems. If that is the nature of C.A.C. in the past and at present, the future nature might well be very different indeed: almost all of the developments now in train indicate a trend towards the easing of individual choice and freedom from the tyranny of map sheet lines and printing schedules.

Even if it is many years before we obtain all of our maps through a television set linked over telephone lines to remote data bases, it does seem inevitable that more and more use will be made of C.A.C. To my mind, maturity will have been reached in this frequently troublesome and noisy sub-discipline when the utility of the computer is taken for granted and the compilation of a volume such as this one is no longer necessary.

REFERENCES

Ager, J. (1977). 'Maps and propaganda', *Bulletin of the Society of University of Cartographers*, **11**(1), 1–15.
Bertin, J. (1973). *Semiologie Graphique*, 2nd ed. Gauthiers-Villars, Paris.
Bickmore, D. P., and Boyle, A. R. (1964). 'An automated system of cartography', *Tech. Symp Int. Cartogr. Assoc., Edinburgh.*
Bickmore, D. P. (1979). This publication, Chapter 11.
C.R.U. (1979). *People in Britain—a Census Atlas*. Her Majesty's Stationery Office, London.
Dale, P. F. (1978) 'The provision of land information' *Chartered Surveyor; Land, Hydrographic and Minerals, Quarterly Supplement* **5**(2), 23.
Dobrin, M. B. (1952). *Introduction to Geophysical Prospecting* p. 336. (McGraw Hill, New York).
Dobson, M. W. (1973). 'Choropleth maps without class intervals?: a comment', *Geogrl. Analysis*, **5**(4), 358–360.
Döös, B. R., and Eaton, M. A. (1957). 'Upper air analysis over ocean areas', *Tellus*, **9**, 184–194.
Gregory, D. (1978). *Ideology, Science and Human Geography*. Hutchinson, London.
I.G.S. (1978) *Regional geochemical atlas: Orkney* Institute of Geological Sciences, London.
Inst. of Met., Univ. of Stockholm (1954). 'Results of forecasting with the barometric model on an electric computer (BESK), *Tellus*, **6**, 139–149.
Meyer, M. A., Broome, F. R., and Schweitzer R. H. (1975). 'Colour statistical mapping by the US Bureau of Census', *Am. Cartogr.*, **2**, 100–117.
Perring, F. H., and Walters, S. M. (eds.) (1962). *Atlas of the British Flora*. London, Thomas Nelson & Sons.
Pfrommer, W. L. (1975). 'Cost and benefit considerations concerning computer-aided cartographic systems' *In* Wilford-Brickwood, J. M., Bertrand, R., and van Zuylen, L. (eds.), *Automation in cartography*, International Cartographic Association, I.T.C. Enschede, 313–320.
Rhind, D. W. (1971). 'Automated contouring—an empirical evaluation of some differing techniques', *Cartographic Journal*, **8**(2), 145–158.
Rhind, D. W., Visvalingam, M., Perry, B., and Evans, I. S. (1976). 'People mapped by laser beam', *Geographical Magazine*, **XLIX**(3), 148–154.
Robinson, A., and Petchennik, B. B. (1975). 'The map as a communication system', *Cartographic Journal*, **12**(1), 7–15.
Robinson, A., Sale, R., and Morrison, J. (1978). *Elements of Cartography*, 3rd edn. Wiley.
Sawyer, J. S. (1960). 'Graphical output from computers and the production of numerically forecast or analysed synoptic charts', *Met. Mag.*, **89**, 187–190.
Schmidt, A. H., and Zafft, W. A. (1975). 'Programs of the Harvard University Laboratory for computer graphics and spatial analysis', in Davis, J. C. and McCullagh, M. C. *Display, and Analysis of Spatial Data*, Wiley, London.
Smith, W. (1978). 'Ordnance Survey AD 2000', *Geographical Magazine*, **L**(6), 361–364.
Simpson, S. N. (1954). 'Least squares polynomial fitting to gravitation data and density plotting by digital computers', *Geophysics*, **19**, 250–257.
Southard, R. B. (1978). 'Development of a digital cartographic capability in the National Mapping Program'. (Paper given to IX International Conference on Cartography, Maryland, July/August 1978.)
Statistics Canada (1979). *1976 Census of Canada. Agriculture: Graphic Presentation*. Ministry of Supply and Services, Canada, Ottawa.
Taylor, P. J. (1976). 'An interpretation of the quantification debate in British Geography', *Trans. Inst. Brit. Geogr.* (NS), **1**(2), 129–142.

Thompson, C. J. (1978). 'Digital mapping in the Ordnance Survey 1968–1978', Paper presented to the ISP Commission 4 symposium on 'New Technology for Mapping', October 2–6, 1978 in Ottawa.

Tobler, W. R. (1973). 'Choropleth maps without class intervals?' *Geogrl. Analysis*, **5**(4), 262–265.

Williams, R. L. (1976). 'The misuse of area in mapping census-type numbers', *Historical Methods Newsletter*, **9**(4), 213–216.

Youngman, C. E. (1978). 'A syntactical structure for map description', Mimeograph.

The Computer in Contemporary Cartography
Edited by D. R. F. Taylor
© 1980 John Wiley & Sons Ltd

Chapter 4
Development in Equipment and Techniques

A. R. BOYLE

INTRODUCTION

This chapter is concerned with the devices and systems necessary to put cartographic data into digital form, handle it in computers, and, if required, output it in drawn form. While the end-point of cartography was traditionally the drawn map, cartographic data now have many more uses in geographic information systems and in computation, particularly in such areas as civil engineering. Today, the drawn map may not even be on paper, but ephemerally on a cathode ray tube (C.R.T.) display screen. Newcomers to the field should consider carefully the requirements they wish to meet, and the techniques and equipment used. Entering large amounts of data can be expensive and with inadequate quality control, incorrect source documents, faulty methods of storage, poor format structure or data description, or any one of these, the end product can turn out to be completely useless.

As a corollary to the above, consideration must be given by a professional cartographer to the needs of computer-assisted cartography (C.A.C.), even if no work in that area is envisaged at the time. An inappropriate method of drafting can change a digitization in the future from a simple, to an almost impossible task. Costs can be ten to hundreds of times higher, and even complete re-drafting may be required.

There has been a major change in approach to C.A.C. since 1976. As the demand for digital data has increased, topographic mapping centres have realized their responsibilities in this new area. Preliminary approaches often involved exorbitant costs and questionable quality of output data leading to immediate problems in the design of storage systems, very high-speed editing stations, and in very high-speed, but also high-quality automatic drafting. When data speeds were low, the slow x–y mechanisms were adequate for drafting but this is no longer the case.

DATA CAPTURE

Much data useful to C.A.C. and geographic information systems are already available in the form of drawn maps. For many years, however, it has only been possible to convert these data into a compatible digital form by tedious, costly and error-prone methods.

The first major step forward was made soon after 1960 by the development of the free-cursor manual digitizing table. This proved to be very good for selected point digitization, but even early work showed it to be inadequate for very large numbers of long, irregular lines. Improvements were made during 1965–70 by operating on-line to minicomputers, and by using etched plates and sheets as tracing aids.

However, it was not possible to consider such a process for the digitization of a complete map series which might consist of as many as 50,000 maps; the time and costs required were prohibitive. Several years of flirtations with automatic line followers (A.L.F.s) took place, simulating the eye and hand of the manual operator. Research since 1976, however, has shown that raster scanning techniques can be superior, giving high-quality result at low cost, providing there is access to the separation transparencies used for colour printing. Obviously everything depends upon efficient scanners, programs, and organization, and these and their combination are rare.

In recent experiments it has appeared that even the separation transparencies are sometimes not clear enough for reliable and most economic automatic digitization. Methods have, however, now been devised, using normal contact printing processes, to satisfy the requirements of the task.

The manual digitizing table is likely to continue to be useful for point digitization, for complex symbology, and for sheets with a very small amount of data. However, the recently attainable cost and quality advantages of scan digitization would seem to tip the balance to that method, providing service bureau facilities can be made available for the 'small' user. The total system cost is high and must be fully used to be economic.

Automatic scan digitization can provide very high-precision and quality line data but does not, except to a limited extent, label the line. The use of computer-aided and interactive display labelling must be considered part of digitizing and consequently this is described later.

Digitizing Systems

Manual Digitizing Tables

For cartography these tables are usually of large physical dimensions (at least 1 m square) and have precision capabilities of between 0.001 in and 0.01 in (0.025–0.25 mm). The difference from 'tablets' is ill-defined, although the latter tend to be of lower precision and smaller. Until recently, there were two main divisions of such tables: those that used an electromechanical follower to track the cursor when pointed or moved by hand, and those that detected the position of the cursor by a system of grid wires embedded in the table surface. However, with the demise of Instronics and their Gradicon digitizer, the former type is now probably unobtainable and, although it served the cartographic community well for many years, it is unlikely to appear again due to higher construction costs and greater maintenance requirements.

Tables now generally consist of a flat plastic board about ¼–½ in in which are embedded or printed a grid of wires. There are various methods of detecting the cursor position relative to the wires and then outputing the position to the system. Most methods depend on an analog interpolation between the wires, which may be spaced at intervals of ½ in or more (1 cm), to give a precision of about 0.004 in (0.1 mm). Linearity depends only on the original mechanical properties of the grid assembly; this is usually good, but it is also possible to compute out remaining errors if it should be required.

Differences between various types of tables are more apparent than real from the user's point of view. One type (Bendix) suffers from an irretrievable loss of output if the operator lifts the cursor from the table surface. Most can be fitted on request with back lighting facilities. Most tables are supplied with cursors which consist of plastic discs or lenses with engraved crosses or circles at their centres. While these are good for locational digitization, they are poor for line tracing, and critical cartographers usually obtain one or other of the scriber point cursors. These are available on special order from most suppliers.

Tables can be operated entirely off-line; the recording of the data output is then usually on to low-cost incremental magnetic tape drives using industry-compatible ½ in wide magnetic tape. Some of these units now include a useful microprocessor to aid in formatting the addition of labels. For locational digitization (and in these days of automatic line digitization this is likely to be the major interest) such off-line operation is adequate from an operational point of view. However, weakness often lies in the low-cost magnetic tape drive, and usually much greater reliability is obtained by interfacing to a minicomputer and using the standard peripheral storage units of that system. Even multiple digitizers can be time-shared onto one minicomputer and still be maintained as a 'background' operation while other work is being done: manual locational digitization is essentially a very slow operation. Many suppliers can now also supply software for such time-sharing, although the programs are not difficult to write.

Much of the complexity of manual digitization was, until recently, concerned with line digitization. Such work is tedious and time consuming for the operator and on-line checking of his work by an attached minicomputer has been helpful. The use of etched plates to guide the point of the cursor and recent developments in error-detecting cursors showed promise. However, with the advent of automatic line scan digitization, this process loses its importance and need not be considered at this time in all of its previously interesting detail.

The addition of facilities to off-line tables for such aspects as scaling are a waste of time and effort unless the only output is to be on a typewriter. This complexity usually reduces reliability and can confuse the operator. When data are to be digitized it is normally because they are to be later manipulated in a computer, and such operations as scaling are much better done at that time.

There has been continuous argument about the usefulness of displays showing the operator the work he has already completed. In general, it is wasteful to do this in a simple way. Faulty lines and erasures are then not usually removed and remain to clutter the image. Screens are usually too small to use without 'windowing'. The best way seems to be to allow the operator to transfer rapidly and at will to a central display station which has full windowing, manipulation, and editing facilities. Detailed examination and correction can then be carried out, but such use is only periodic in a well-run shop. The process of adding such a fully interactive capability to each table adds costs and can be regarded in most cases as 'overkill'! One good argument for an on-line display of a simple nature has been to show the operator what has been digitized on a previous run and allow him to make an addition or modification to fit as closely as possible, avoiding much tedious later work of editing such joins. However, this work usually requires a real-time 'scale and rotate' program which uses considerable computer time, delays the operator and makes the display complex once more. It is an important philosophy to make the operator, when digitizing, concentrate closely on the work in hand and not have his eyes diverted to other devices such as displays or even keyboards. More will be said about this later.

The future use of manual digitizers is mainly in point digitization rather than for line streams. Point digitization and character and symbol recognition are more difficult to automate than lines, and the discrimination of the operator on point data is useful whereas it tends to lead to errors on lines. One aspect that should be mentioned, however, is that large-scale urban maps having very long straight lines and simple curves may well be handled by point digitization with the operator selecting the most useful line specification points. This policy has been pursued with appreciable success at the Ordnance Survey in the U.K.

Automatic Digitizers

In automatic line digitization, the scanner now appears to be the best possibility. Automatic line followers were attempted, but as they resulted in general failure they need not be considered fully here. The problems of automatic scanning are partly in a good design of scanner, but more in the conversion of raster scan data to vector form, at present being demanded by cartographers.

Scanning

Automatic line followers essentially replaced the human movement of the cursor on manual digitizing tables by mechanical means, a position sensor being used instead of the operator's eye. However, the large mechanical inertias meant that line followers of this type were very slow and, as the equipment was expensive, costly because of the slowness. They also suffered from the necessity for operator intervention to start the follower on each line; this problem even inhibited the much lower inertia followers using laser beams (e.g., Sweepnik, first developed in the early

1960s for bubble chamber track digitizing in nuclear work) and would thus not operate efficiently or economically.

This method at first appeared attractive to cartographers as it output a string of coordinates for a line. However, the disadvantages led to the more indirect method of scanning, followed by software line following as a preferential method.

In scanning, the map separation sheet is automatically examined strip by strip, each strip being only 0.004 in (0.1 mm) wide in a typical arrangement. Cartographic lines appear as position dots (or pixels) separated from adjacent line pixels by the pixels of other lines. The computer, knowing the code of the arrangement, has to reorder these to form the desired series of coordinates down the line. This rectangular coverage of a sheet strip by strip is called a raster scan and, except for the different format sizes and resolutions, is very similar to that used to make up a normal television image.

As one would expect, there are intermediate stages possible between the two extremes described above. One consists of scanning with a laser beam until a line is detected and then changing to a line follower in its movement; the other is to convert the raster to vector line dot by dot as the scanning proceeds. Both of these methods cause very much longer usage of very expensive machinery and seem also to lead to an accumulation of unwanted errors in the attempts made so far. It now seems preferable and 'cleaner' to scan as rapidly as the data can be accepted into some storage system and then subsequently and separately convert the raster scan to vector line in a separate computer (large main frame or minicomputer, as wished).

At this time, scanners are typically drum systems similar to large-scale facsimile units which are now widely used in the graphic arts industry. In general they take approximately one hour to scan a sheet at 250 lines per inch and costs about $100,000, to which must be added the cost of a minicomputer controller. However, a time of one hour for production of one sheet which may, in fact, have few lines, is long and thus costly. It is expected that shortly these drum scanners will be replaced by flatbed laser units able to scan a sheet in a few minutes.

The problem of the scan to vector conversion has been a major problem for a number of years. Why this should have been so is not clear, because it has now been shown that efficient conversions can be run at low cost and in only a few seconds or minutes. It is possible to do this even on a minicomputer and one certainly does not need the major main frame that some establishments have suggested.

The output from scanners can also be used for optical character recognition and analysis of numerics, alphanumerics and symbols. It is not usually suggested, however, that this has other than very specialized use, for example, in the reading of numerical depth information from hydrographic charts. This method for 'name' recognition is not recommended as that can be done better by other means such as key-punching gazetteers and then moving the names to their exact position by interactive editing once the lines are available for positional background.

One of the shortcomings of automatic line digitization is that the lines are obtained merely as unlabelled 'spaghetti'. Advances are now being made in the use of program de-symbolization to produce a centre data line from symbolized map lines; at the same time as doing this the labels can often be created automatically. Progress is also being made in program assistance for labelling, such as the annotation of intermediate contour values once index ones have been found. The numerics available on some contour maps are not usually directly useful in automatic operation, and at least some of the contours must have their values added by interactive editing.

It will be seen, therefore, that a modern digitization system for existing map separations will consist of:

(i) scanner;
(ii) scan to vector conversion program;
(iii) manual table for numerics and locational symbols etc.;
(iv) interactive display and edit station;
(v) Programs to aid in labelling.

Of the above, the first two are probably best done by a service bureau operation. The last three are better done in-house, as the knowledge of the user is critical for efficient results.

For good automatic operation, the map separations must be properly prepared and controlled in quality. Generally only some opaquing of 'impossible' situations is required, as far as manual intervention is concerned. Preferably, the system must be run on a serial planned basis 24 hours a day, if good quality and output are to be maintained. Periodic use, often by various persons, can lead to trouble.

The lines to be digitized must be of correct density (or clarity if a negative transparency is required for the system) and with well-defined edges. With early programs only a single width with a very close dimensional parameter could be accepted (for example, Canadian Geographic Information System (C.G.I.S.), Ottawa) but now widths over at least a 4 : 1 ratio can be used. This 4 : 1 ratio is usually conveniently set by scanner resolution and spot size to a range of 0.002–0.008 in (0.05–0.2 mm).

Some conversion programs can record the line width found; others do not do this as such processes are time-consuming. Some programs have trouble if lines are wider, then creating multiple unwanted nodes, while others only create an incorrect data line centre. Generally lines thinner than the specified range result in broken line data. This can also happen if faint or pencil lines are attempted.

Symbolization of dashed and dotted lines is not usually a problem if the width of such components of the line is within the range, but combinations of symbols and lines, and particularly if the symbols are 'blobs', are better left to manual digitization (often these are mainly straight lines which can be handled easily). Side dashes, such as for depression and fill ticks, and for telephone posts at the side of a road, can be handled usefully by the de-symbolization routines.

While it is possible to remove all unwanted data by opaquing, these could also be retained through the digitization process and removed on the interactive display and edit station. Sometimes one is best and sometimes the other. A routine way of doing this, which can save a large amount of work, is to carry out, in the preparation of the sheet, a special contact process involving line widening and thinning. In any case, copies have to be made for scanning; the special operation described below involves only a little extra effort when production is continuous. The process is as follows:

(i) Contact print positive to negative transparency using a 0.007 in (approximately 0.02 mm) transparent spacer between the emulsions, and a diffused light source. The light bleeds around the line edges and, as a result, completely 'removes' all thin lines (say 0.004 in (0.1 mm) and less).
(ii) Repeat the process and contact-print this negative to a positive transparency. By using a little wider spacer (line thickening now occurs due to light bleeding out *through* the line slot), a copy of 'thick lines only' can be obtained with these lines slightly larger than they were originally.
(iii) Use this positive transparency as a mask on the original negative and contact-print to a positive transparency without spacer and with a point light source. This process will produce a 'thin line only' sheet very suitable for automatic digitization. All thick lines and 'blob' symbols will have been removed and these can be entered by manual digitization. The process is very good to remove 'house' symbols and highly symbolized lines on culture sheets, and to remove numerics from contours.

Audio Input

It has already been mentioned that the operator's attention should not be diverted to other than his cartographic working sheet. Many operators, not having the training of typists, find that a keyboard causes slowness, loss of concentration and thus probably of data, as well as actual errors. A method now finding favourable acceptance, particularly in view of rapidly reducing costs, involves the use of a microphone and audio input directly to the control computer. Systems to handle fifty or more individual words are relatively easy to obtain and use, and reliable in performance.

The audio method is particularly good if an audio output device (also low-cost) is added to the computer to provide an audio feedback directly and immediately to the operator. Without appreciable thought, the habit can be gained of an automatic 'mental loop'; if an incorrect answer be given, the operator calls 'no' and repeats. Continuous faults probably indicate the necessity for a replacement of the stored template of the word in the computer memory.

This method is now also finding acceptance for photogrammetic work and for display and edit stations.

Photogrammetric Input

It is impossible to discuss cartographic digitization without also describing photogrammetric input. Methods vary with the aims of the establishment, the map scale being used, and the terrain under analysis. Small-scale (25,000 to one million) users tend to prefer the present manual drafting processes followed by automatic digitization, whereas users preparing large-scale engineering drawings prefer on-line digitial output, possibly with individual display and edit units for data examination and correction as the work proceeds.

Establishments are now commencing the direct automatic digitization from stereo models to produce D.T.M.s (digital terrain models) and contours. The work is very promising and is further described by Harris in Chapter 5 of this volume. The parallel preparation of an orthophoto map, which often takes place, also affects the way in which the data can be prepared.

A method seen by the writer and which recommends itself for scales smaller than 1/20,000 is as follows:

One operator has available to him the orthophotomap of the area and also the necessary models on a B8 photogrammetric reader or similar unit. He background-prints the orthophotobase onto a number of sheets and then, wherever possible, uses direct scribing to obtain the final line work; this consists mainly of roads and drains. When he cannot see the necessary detail he refers to the stereo model. Contours, he obtains from the B8 first as ballpoint copy and then scribes these lines, correcting as he goes by referring to the orthophoto background and other information.

The sheets produced are then sent for digitization prior to the addition of alphanumerics and most symbols (for example houses) making automatic scanning straightforward.

It has recently been suggested that the operator should avoid the drafting of symbolized lines and only scribe them as full 0.004 in (0.1 mm) wide lines. These would be scan digitized and sent to automatic drafting, at which time programs for symbolization would convert the pure line centre data to a symbolized form ready for colour printing.

DATA MANIPULATION

The data from digitization are not directly useful to cartographers. These are in machine coordinates and must be transformed to a geographic base. There are also other batch processes which must be carried out.

However, the most importance part of data manipulation is interactive display and edit, which allows the cartographer to examine and modify data as he wishes. Without such a system he would be operating blindly. While he might well contract out digitization and precision drafting his display and edit must be in-house.

Data manipulation is best divided into two parts. The first is preparation for storage and the second for drafting. If the storage of data in digital form is the end-product, a data base is designed in one manner; if that data may also be used for drafting of revisions and new maps, then some data may best be stored in two forms. For example, the centre line of the road may be best for geographic modelling, but the road casings are needed for revisions.

The manipulation processes depend to some extent on the storage methods used. Some of the latest, especially those useful for cartography, will be described.

Batch Processes in Data Manipulation

Digitization to Storage

(i) If scanning is used for input it is usually necessary to convert from scan to vector line form in C.A.C. It should be remembered, however, that this is not essential, and area data such as in forestry may be better stored and handled in that form.

(ii) Line data from digitization must be compacted. This process is usually called 'weeding' or 'culling'. For example, straight lines from automatic digitization will be in the same form as irregular lines, as a series of increments; these must be changed to a simple specification of two end-points.

(iii) There are many symbolized lines in dot and dash form on drawn maps. These lines must be coalesced to form complete feature lines.

(iv) As a result of de-symbolization and other known factors about a separation, some lines can be labelled automatically. Some require interactive aids.

(v) Certain types of 'short-line' data can be handled by special programs. For example, fill and depression ticks should be removed from the contour lines and stored separately.

(vi) Data in digitizer machine coordinates must be transformed to geographic ones of the data base. This may be a simple conformal transformation or a complex one to longitude–latitude. While mathematical formulae may be used for small scales of less than 1/100,000, mosaicing methods must be used at larger scales.

(vii) In certain applications, it is useful to form polygons automatically from the node and line data files.

Storage to Drafting

(i) Data must be selected in type and in area, on entry to the work area.

(ii) Data must be transformed into the map projection and scale required with the appropriate warnings to the operator if the resolution of the data at any position is inadequate. The end product is drafting machine coordinates.

(iii) Data must be symbolized as required.

(iv) Data must be smoothed to meet specifications at the scale required.

(v) As methodology improves a degree of cartographic generalization should be done as scale is reduced.

Interactive Display and Edit Processes in Data Manipulation

While as much work as possible should be done by batch processing, such work must always be checked by a cartographer and his experience will be essential for certain manipulations. The tool for this task is an interactive display and edit unit.

These stations consist, in the most useful operational cases, of a small but complete minicomputer system and include display, keyboard, medium capacity disk and interface link. In general, because of the very large amounts of data, time-sharing onto a central disk system via standard computer links, is not feasible; the access time delays can become infuriating to the cartographer.

For display, it is essential to be able to select data and display and 'window' it quickly and easily. Editing requires very fast access to any location or line pointed to, and the possibility of performing any of a flexible group of functions on the location or line selected.

Interactive display and edit is required at three stages. Two of these are the same as for batch work, that is, digitization to storage and storage to drafting. The third is to examine and manipulate the data in the data bank itself. A general description of interactive display and edit will be given followed by a consideration of the special problems of each of the three stages.

Unfortunately, a close examination of the various self-contained minicomputer systems offered for this work shows that many are inadequate in performance compared with the potential of a properly designed system. This particularly applies to speed of operation, ease of use, and flexibility of functions.

If speed is to be maintained, and this is the essence of good interaction, special software structures and sometimes special hardware must be used.

Poor system design in some existing systems has resulted in a poor speed performance. Frequently this can be improved ten or one hundred times. Manufacturers' interfaces to displays have been tuned more to telephone line connections than to direct output from the computer bus. Strangely, many computer scientists working on slow line communication terminals have come to accept slow speeds. While a cartographer might be impressed with the images produced in the first few days of working with such systems, he may soon find himself wishing for his fast, easy to use, flexible, pencil and paper!

Cartography is basically different from other interactive display systems such as those for engineering or circuit design, or even from geographic information systems which may include relatively small amounts of low-resolution map data. It is different in that almost every resolution element has to be specified in position to handle the appreciable irregularities of many cartographic lines. When this is taken into account, it may mean that 10^6 coordinates have to be specified for a single separation. Only very large-scale urban maps can be approximated to engineering drawings in content and bulk of data, where only line end points are specified. Another difference is the much wider range of editing facilities required.

Modern cartographic data structures try to limit the number of coordinates by weeding processes, in which data points are only recorded when the line moves away from a specified line situation by more than one digital resolution element. In the simplest and most usual form, the specification is a linear one, but most structures can be used for either.

The problem is that in reducing the amount of data to be stored, handling time is increased. In many cases the increase can be swallowed up in other I/O (input/output) operations, but in interactive display and edit it can slow operations to the nuisance level. This nuisance level must be considered from a production cartographer's point of view. In the writer's opinion, this is when times of access of an image or of modification are longer than about 6 sec. On the other hand, trying to obtain times of less than 2 sec is not significant. It will be seen that the time range is relatively narrow.

The major problem, therefore, in interactive display and edit operations, is the handling of literally millions of coordinates at very high speed, while the operations on each coordinate are almost trivial. The desired results are best obtained by the use of a dedicated minicomputer at each station fitted with its own display, keyboard, and disk. The minicomputer can be small (and as microprocessors become faster they should be adequate) and every cycle is critical.

Another problem with interactive display and edit is a psychological one in that the operator is examining an image in detail at close range. Considerable mental stress can be caused by 'refresh' displays which are, of course, adequate for normal television viewing. This is so even when they are refreshed at 60 Hz, instead of the 30 Hz which is quite unacceptable for map work, although generally satisfactory for purely alphanumeric use. For this reason, the storage type display (such as Hughes or Tektronix) has been more frequently used; they are also advantageous in resolution and cost.

As far as cost is concerned, the price of a storage display at about $20,000 has to be compared with that of a lower-resolution 'refresh' display with at least four image plane memories to give local storage. Only with very limited data can the minicomputer core be used for 'refresh' without excessive image flicker. One of the few advantages of 'refresh' is that it is at this time almost the only way of obtaining colour. If colour is an essential, then the flicker must be accepted, although the generally greater 'softness' of a colour image tends to help.

Interactive Display and Edit—Digitized Input Data

This is the interactive display and edit area which is relatively the simplest and has received most attention from suppliers. In the main, the systems have been obtained to attempt to correct the very large numbers of errors introduced by manual digitization; these are usually concerned with undershooting and overlapping of lines at junctions.

In view of the movement towards automatic scan digitization, such errors become unimportant as they are unlikely to occur. The main shift of emphasis is towards labelling of the data, as well as erasure of unwanted data which may have been digitized by mistake as lines—for example, alphanumerics and symbols. The process required is to point to the line or location and key or 'audio' in the label or correction.

The main problem is that with automatic digitization a very large amount of data will be present, often up to 10^6 coordinates from a single sheet. While many systems can handle up to 10^4 with little fuss, the extrapolation can cause excessive time delays. The structure of the data, the storage method and, most importantly, the structure of the data directory, must all be evolved in great detail. One system on the market at this time has no directory system at all and relies for its 'speed' on fast passage through all the data each time a request is made. A good directory can confine much of the work to itself alone.

In general, a good directory system will contain a data line reference number, a bounding rectangle to define the area for fast search, and a 'part' label. It may possibly contain line start and end coordinates and line length. There will be room for various tags such as that for 'delete'.

It is not necessary that structures for display manipulation are the same as those used for storage and it may be preferable to modify and expand the directory during input to the display operation. This may simplify operations and speed work appreciably.

Interactive Display and Edit—Storage to Drafting

This is a very different problem from input digitized data display and edit. Many more operational features are required, and only one manufacturer to date has attempted this type of work. In fact, to be truthful, the need has not been there because of the difficulty and expense of obtaining a sufficient amount of digitized input.

In display, the full map symbolization facilities must be available for at least line width as well as locational symbology. It may also be necessary to symbolize areas by angled lines, fill-ins, or even colour. It must be possible to symbolize either from input data specification or by operator intervention and, in the latter case, the code selected by the operator must be recorded. This coding has to be either as 'permanent' or 'temporary'.

Moreover, the line widths and symbol sizes must be automatically increased or reduced as windowing is carried out, as the operator will be appreciably concerned with interference between one line or symbol and another.

For selection of data for drafting, the operator must be able to 'tag' data. The system must be able to select these tagged items and output the symbolization codes to the drafting system.

In editing, it must be possible to delete and move symbols and names, and even individual letters within names. Movement must be on the basis of altering the reference position or moving to an offset for clarity of drafting. It must be possible to alter the shape of lines to give the design, clarity, and generalization desired.

It is particularly in the manipulation of symbolized line data and names that major program additions are needed. These are difficult as they must be carried out at very high speed to avoid operator annoyance. The bulk of the data is not usually a major problem (although one to be considered) as lines have usually had their data compacted and only one map area is under consideration at one time. A problem only arises if a small-scale map is being made from large-scale data.

As in all other display and edit functions, the directory system is of vital importance.

Interactive Display and Edit—Data Bank

The use of an interactive display and edit system for handling the data bank itself is generally similar to that for digitized input data, but has to accommodate an even larger bulk of data. In fact, the amount of data can be so large that either the longer access times have to be accepted or a subsection of the data be transferred to the local disk for operator use.

The future for this task lies in the application of laser-written optical disks. The costs of a playback unit should be relatively low (less than $10,000) and thus can be fitted to the display and edit unit itself. The disks hold a very large amount of data equivalent to 10^{10} bits (reliably) or, with good packing, about 1000 topographic maps including all separations and content. It will thus be simple to transfer a copy of the disk to the display station, by hand. Any other way appears to be nearly impossible as computer links would cause excessive slowness in display manipulation or delays if batches of data were transferred. The optical disk is permanent and cannot be directly modified: when editing is to be done it is, in fact, carried out on a magnetic disk working in cooperation with the optical disk. Periodically new optical disks will be made from the composite. The liaison consists of transferring the directory from the optical disk to a magnetic disk unit, and tagging that for deletions or modifications and adding new data.

The directory itself is, of course, very large and must be structured for efficiency in transfer and to ensure that only the functional section is in use at any one time.

With this very large amount of data to be handled, the data and directory structure becomes of vital importance if speed is to be maintained.

The types of task to be carried out are as follows:

(i) visual examination only;
(ii) deletion of faulty data and possible replacement;
(iii) tagging data with a reliability or usefulness value (generally when multiple data become available for an area);
(iv) general label and coding changes;
(v) correction of line data and locational position when new information becomes available.

Full facilities must be available in the data bank itself for subsequent incorporation of editing changes made. Otherwise the work would be useless.

Time-Sharing Aspects of Interactive Display and Edit

If a large amount of cartographic data with irregular line form is to be handled, time-sharing cannot be considered as it will usually cause excessive slowness and annoyance to the user.

Time-sharing is appropriate in special cases such as for locational data only or for large-scale urban sheets and engineering plans that can be described with a relatively small number of locations; that is, long straight lines by their end points.

The time requirements are such that a single dedicated minicomputer—even a small one—can meet the needs if local disk storage is also available. It has to be appreciated, however, that a 'large' minicomputer is not faster than time-sharing except for calculations; in cartographic data handling there are few calculations other than of a trivial nature. Cycle time, and the usage and availability of each cycle, are matters of importance.

Display Screens in Development (including large screens)

As described earlier, the bulk of cartographic work has been carried out by storage type screens. The Tektronix 4014 (19 in diagonal) and the Hughes unit are predominant.

Because of the large numbers of vectors, computer-core refreshed screens are usually impossible to use, except for large-scale urban areas and engineering plans. The refresh screen also causes psychological problems for the viewer who has to examine the data in detail at close range. This latter aspect restricts even the use of refresh displays with self-contained memories.

In general for topographic work, the computer must supply each point, pixel by pixel; vector generation by hardware is only marginally useful. Hardware alphanumeric generators may be used with advantage but they are often too limited in vocabulary. In general each pixel must be addressed individually by the computer and the hardware interface must be able to operate at an adequate speed of less than 100 μsec per point.

Cartographers would always like to have larger screens—full map size—but these will still take some years to develop. Plasma and liquid crystal screens are the most likely contenders but these will be expensive given the size and the level of detail necessary. In practice, cartographers generally find a 10 in diagonal screen is adequate. A particular advantage can be obtained by the simultaneous use of a pair of these, one giving a reduced-size full image and the other the enlarged working area.

One complaint about storage screens is the fact that, at least on the Tektronix, individual erasures cannot be made. The whole screen must be erased and regenerated without the unwanted item. However, with good planning and a good directory and control system, this can usually be accomplished within 2 sec and this is acceptable. Some systems on the market, however, are incredibly slow in operation.

Pointing to position on the screen can be done by roller ball, joystick or 'mouse'. The latter is a small box contained in the hand and moved over any surface. The two orthogonal wheels beneath the box provide x and y movement information to the cursor or spot on the screen. In general, light pens are not useful or used for cartographic applications, as they are too crude in positioning and the operator's hand hides much of the image.

Some users like and need colour. Unfortunately, no colour storage display units have yet been made available, so that refresh types must be used. One possibility is the beam penetration tube made by C.S.F. France, but the standard television unit is still the most easily obtained. However, as this only has low resolution, the tasks for which it is used have to be carefully selected. A new advance in use has been to apply large-screen projection type television screens for geographic information system requirements. The cartographic detail required is relatively low, but the image is acceptable to viewers.

If used for more detailed cartographic work, a smaller map area at greater enlargement must generally be used. However, the colour refresh display with its softer gradations does not seem to cause similar flicker problems as those caused by purely black and white versions. One particular use is in the combination of cartographic data with Landsat imagery, where colour discrimination is a major advantage.

STORAGE METHODS IN DEVELOPMENT

If reliance had to be placed in the future on magnetic tape disks and core for high-speed operation, the picture would indeed be bleak when cartographic production is considered. Fortunately, it does appear that laser-written optical disks will soon become available. These disks can accommodate up to 10^{10} bits of data in a reliable archival storage mode. The cost of playback units is expected to be low, so that multiple units can be available, fitted to each display and edit station. It is usually accepted that one complete topographic map can be compacted into 10^7 bits, so that the capacity of the disk is 1000 topographic maps in full detail.

Optical disk recorders will also be available but, as appreciable care must be taken and the work would be infrequent, it is probably best that this be handled by a service bureau operation. Disks can also be duplicated at low cost (probably about $20).

The recording is done by a fine laser beam causing minute eruptions of about 1.5 μm in diameter on the metal surface of the disk. These eruptions are patterned to remove errors in the case of a surface blemish. Readback is done, either by another laser, or by a simple optical system. Special servo controls are added to allow for lack of disk concentricity and flatness.

It does not appear at this time that holographic storage methods will be useful.

The optical disk is always used by the cartographer in conjunction with a magnetic disk to carry a working copy of the directory and to record updates. Periodically a new optical disk is made to include the changes made.

DRAFTING SYSTEMS

For many years the Gerber 32 and Kingmatic x–y plotters have been the mainstay of precision drafting. However, they are slow and times of more than 30 h for one map sheet are not uncommon, although the resulting line and symbol data are superb in quality.

In general these plotters use optical heads, writing with light on litho film, rather than pens or scribers. These are expensive optical devices with light source, lens system, symbol disk and a number of mechanical control mechanisms. They are usually arranged to use an air float support to maintain exact focusing distance from the emulsion.

These optical heads have only a limited number of symbols, as low as 24 in some and 94 in the largest; changing symbol disks in darkroom conditions in the middle of a plot is not easy. A method of using a C.R.T. light head has been devised and appears successful and convenient. It can be used to produce single symbols and alphanumerics at large size on the C.R.T. screen and these are then imaged down to required size on the emulsion. The symbol library is 'unlimited', as it is stored in digitial form on magnetic disk. Rotation and size are calculated in the control minicomputer. The unit can also be used to draw complicated symbolized lines by moving a rectangle shape in the line direction, stopping periodically to create a locational symbol.

A method of using a very high precision C.R.T. has also been developed to expose complete rows of alphanumerics at a time and allow for fast name placement. The extra weight and cost of this system, however, hardly justifies the extra speed obtained over single-letter operation with a simple unit.

The same precision unit has also been used to expose complete map areas of about 2×2 in (5×5 cm), to allow a precision drafting table to be used for 'quick look' high-speed operation. Again, the cost is not likely to warrant general use.

Special higher speed, but lower precision, x–y plotters are also available such as the XYnetics magnetic float unit. These can be used to advantage in some cases, but the more usual unit for quick-look is the drum plotter, of which the Calcomp is perhaps the best known. Other high-precision, but yet still high-speed, units are available from companies such as Gerber. Generally line quality is only medium, as light heads are not normally available with these approaches.

Drum scanners are now widely used in the graphics arts industry, usually operating at about 2000 lines per inch; these give a high-quality image, satisfactory for many map applications. Writing is usually by a finely focused laser beam as the drum rotates beneath it.

The difficulty in application in cartography has generally been in the vector to scan software, although precision drum devices themselves are difficult and expensive to make. Much work still remains to be done in the software area in order to obtain economic and high-quality results.

The latest approach is to use 'flatbed' mechanisms, where the laser beam sweeps from side to side as the film sheet passes below it.

Scanning methods inherently have the advantage of unlimited symbology for locations, lines, and areas. There are great expectations that all cartographic drafting will be done by these precision scanners within about a 3-year period.

ECONOMICS

Most C.A.C. has been carried out by government departments and these are noticeably reticent about giving costs. In fact, claims are even made which turn out to be invalid, but are not corrected in the literature. In order to assist the reader in 'ball park' ideas of cost, the writer gives the following assessment. This is based on experience, a knowledge of the experimenters' work to date, and an appreciation of the next stage of development.

The costs given below only include labour and equipment amortization costs; overheads are neglected. As well as costs, the throughput also has to be considered if efficiency is being assessed. In most cases the writer assumes the use of efficient programs and hardware—a situation not always met in practice!

One of the important costs recently 'discovered' is the high cost of format conversion. More responsible costing systems could make the use of common formats highly desirable.

Equipment
(High-quality, precise devices are assumed).

Manual digitizing tables	$10,000 to $20,000
Interactive display and edit stations (including controller)	$50,000 to $150,000
Minicomputer systems	Approximately £100,000
xy drafting systems (including controller)	Approximately $250,000
Light heads for above	Approximately $50,000
C.R.T. light heads (with controller)	$50,000 to $150,000
Scanners for digitizing or drafting (with controller)	$200,000
Laser-written optical disks drives (read only for fitting to minicomputers)	$10,000

Note: Actual systems costs are difficult to indicate as a number of digitizing tables may work with one minicomputer system. Many units require their own controller but may store their data on an attached minicomputer system.

Costs of Digitization
A topographic quadrangle at 1 : 24,000 (average eight colour separations) is taken as the basis. Obviously this is an average figure as some 'quad' sheets are complex and others simple.

1. *Manual digitization* (from experience)
This results in poor-quality line data and thus there are major costs in editing correction.

	Costs ($)	Times (h)
Manual digitization	5000 per map	i.e. 250
Editing correction	3000 per map	150

Elapsed time for above is about ninety days.

2. *Automatic scan digitization* (from test results)
This results in good-quality line data. Little correction is required but interactive display labelling is needed. Some data have to be manually digitized (e.g. bench marks and spot heights).

	Costs ($)	Times
Sheet preparation	200	30 min
Scanning and scan/line conversion	200	20 min
Interactive display labelling	120	6 h
Manual digitization of locations	120	6 h

3. *Costs of production of laser-written storage disks* (estimates only)
   First disk $200 (storage equivalent to 1000 map sheets)
   Copy disks $20

4. *Costs of output drafting* (estimates only for scanner)

	Costs ($)	Time (h)
(a) *xy* precision drafting	250	10
(b) Scan drafting		
Vector/scan conversion	1000	1
Scanner	200	1

The cost of $1000 for vector/scan conversion is excessive at the present time and must be reduced—development is at hand.

CONCLUSION

Over the last two decades much progress has been made in the development of equipment and techniques. These changes have taken place in a relatively short period of time as the following brief historical calendar of events indicates.

1960		First approaches to C.A.C. Minicomputers were non-existent and even magnetic tape was new.
1961		Free cursor manual digitizer tables were developed and marketed—largely to meet cartographers' needs.
1964	(a)	C.A.C. systems made their first hesitant steps.

	(b)	The first minicomputer was marketed.
	(c)	The direct digitial control Calcomp plotter was marketed.
1965–70	(a)	The first C.A.C. systems started to work. These used manual digitizing tables and large precise x–y plotting mechanisms.
	(b)	Various reasonably large digitization programs were undertaken (such as, World Databank No. 1 by the CIA and the 1/250,000 series of the U.S. by Defence Mapping Agency in the U.S.).
	(c)	The first experiments in interactive display and edit using the IBM 2250 were made.
	(d)	The first experiments in automatic digitization were made (some *intended* to be production systems).
1970–75	(a)	Advanced interactive display and edit stations using dedicated minicomputers were developed and marketed. This caused a fall-off in use of 'quick-look' plotters.
	(b)	Experiments continued in automatic digitization by line-following and scanning.
1975–78	(a)	Automatic scanning became available and efficient (automatic line-following was dropped).
	(b)	Experiments in output drafting by the use of 'graphic arts' scanners were commenced.
	(c)	Experiments were made with laser-written optical disks for the storage of cartographic data.

It is clear that the pace of change is accelerating and that some equipment and techniques have very rapidly become obsolete. The pace of change in this field is likely to continue.

The Computer in Contemporary Cartography
Edited by D. R. F. Taylor
© 1980 John Wiley & Sons Ltd

Chapter 5
The Application of Computer Technology to Topographical Cartography

LEWIS J. HARRIS

TOPOGRAPHIC INFORMATION AND ITS PRESENTATION IN THE TRADITIONAL ERA

INTRODUCTION

Certain characteristics of traditional topographical maps remain in the era of automation and influence production specifications. A clear understanding of these characteristics is required before venturing into automation.

The application of computer technology has revolutionized the processes and procedures of commercial accounting, of industrial inventory, of controlling machines, and of solving scientific problems. Since the early 1960's, large and expanding industries have been formed to make computer equipment, to design computer systems to solve specified problems, and to devise new fields of application of computer technology. No scientific or engineering profession or discipline has been able to ignore the effect of this revolution, and each has been obliged to review its traditional practices, and to assess the relevance of the new technology. The computer industries have not been backward in proposing applications and solutions but, at the same time, the professions have not always been fully aware of the contemporary status and limitations of the relevant computer technology. It is necessary, on the one hand, that the professions define their needs clearly and do not abandon standards evolved and proved over many years of traditional practice; and, on the other hand, that representatives and workers in the computer industry fully appreciate the professional needs. Computer technology is exciting in its theoretical exposition. There may be a possibility of becoming overly excited by the prospects described of believing too readily that the application of computer technology can fulfil all the cartographic requirements, and, thereby, of overlooking the essentials and range of traditional cartography. Therefore, in discussing the application of computer technology to topographical cartography, it is advisable to begin with a description of traditional cartography and of the topographical map, in order to give a background against which the use of computer technology and digital data can be viewed.

Cartography is a graphical method of description and a visual means of communication. It should demonstrate the relationship of the data depicted with a reference cosmographical surface, usually that of the earth or part of the earth's surface. A prime requisite is that the cartographic presentation permits the reader to make, from the graphical synthesis of the features, deductions which are not apparent from the consideration of the indivudual features.

Different countries have different views on the content of the subject of cartography. In Great Britain, the Royal Society defines cartography as 'the art, science, and technology of making maps, together with their study as scientific documents and works of art ...' In some other countries, cartography has been defined as being limited to the depiction of the surveyed data. The influence of the use of computers is to favour the widest definition, due to the making of maps from data in digital form, and to the scope for the mathematical analysis of the digital data. With the computerized automation of cartography, it is impossible to exclude completely from cartography the activities of geodesy and photogrammetry.

Topography in the original Greek means the description of a place, whence a topographical map is a graphical description of a place. Topography, in the modern sense, has also been defined as the detailed mapping or charting of the features of the environment, especially by means of surveying. Topographical cartography can then be taken to mean the cartographic description of a place or environment. In other usages, the term 'topographical map' has conveniently been used to refer to certain scales of maps, say, between the scales of 1/10,000 and 1/250,000, the scales larger than 1/10,000 in such instances being called the engineering and cadastral scales, and the scales smaller than 1/250,000 being called the geographical and atlas scales. In the practice of applying computer technology to cartography, a topographical map should be taken to mean a cartographic description of a palce or environment in terms of the visible features to the extent that the scale of the map permits.

Automation is defined as a sytem of operating a mechanical or productive process by highly automatic means using electronic devices and digital computers.

TRADITIONAL MAPS

Viewing the range of traditional maps will help to place the topographical map in its proper perspective within cartography. Up to the eighteenth century, nautical charts, topographical maps, town and estate plans, route maps, atlases, and world maps comprised the whole of cartography. To these have since been added a large variety of maps and charts to describe themes or to satisfy special requirements: physical, geological, and land-use maps; aeronautical and hydrographic charts; road, street, traffic flow, and route accessibility maps; population and political maps; maps to depict medical statistics and all the statistics of the environment which have a geographical connotation; maps for land registration; and maps for administration and for development.

The topographical map, generally, provides the foundation or the source for the production of the foundation for these thematic and special maps and overlays. At the larger scales, topographical maps are used for administration, development, engineering, and land registration, or as an index to the land registration system of countries. Consequently, adequate and up-to-date coverage of topographic maps is a primary requirement of a country to permit efficient administration, to aid development, and to provide the source for smaller scale maps and the foundation for thematic maps. The topographical maps are usually produced by or under the direction of governments, national, regional, and local.

Certain characteristics of the topographical map series covering a country, region, or locality will require to be kept in mind in applying automation, and these will be described as they apply to the traditional topographical map.

The parent or basic scale is the largest scale of topographical map required to depict the environment adequately, varies from area to area, and should be selected by the mapping organization after considering such factors as the amount and density of the topographical details to be mapped, the extent of the area, the accuracy of positioning the features on the map, the value or potential value of the land to be depicted, and the resources and time available for completion. The largest topographical scale of mapping is usually termed the parent or basic scale. From the parent scale topographical map, the national, regional, or local family of topographical map series can be derived in a decreasing order of smaller scales. The topographical maps will have been projected on a selected projection, usually an orthomorphic or conformal projection, which enables the distance and direction between points and the scale error in each area of the map to be determined. The maps will have been cast on selected sheet lines, either grid or graticule, will normally have a chosen reference grid, and will have been drafted with specified symbols for each feature.

The concepts of the parent scale and the family of scales are well exemplified by the national mapping programme of Great Britain, in which the maps of the major towns are surveyed at the 1/1250 scale; of the minor towns and rural areas at the 1/2500 scale; and of the remaining mountainous areas at the 1/10,000 scale. From these parent scale maps are derived the smaller scale maps to create the national family of maps. Undertaken by one production agency—the Ordnance Survey of Great Britain—the programme is suitably structured for the comprehensive application of computerized automation. However, from the national viewpoint, this concept of a parent scale may not be completely true, because 'islands' of even larger scale maps may have been completed or may be completed in the future by local governments or by various organizations for their own purposes. Such a situation is normal in countries of great extent, where smaller and generalized parent scales provide most of the national coverage. In Canada, the current federal programme aims to provide national parent scale coverage at the 1/50,000 scale. The provincial governments and major municipal governments, will, in due course, undertake any larger scale maps where the requirements justify and the resources permit. These provincial 'islands' of larger scales will become the parent scales of those areas. The federal, provincial, and municipal programmes should be coordinated and this requirement becomes more important with automation.

Systematic revision of all of these maps is necessary to ensure that they are kept up to date with any changes in the topographical scene and with any change of geodetic datum and geodetic control, which will affect the graticule of latitude and longitude and any rectangular grid calculated from the latitude and longitude. Such changes may alter the sheet lines of the maps, that is to say, the boundaries within which each set of topographical features creating the map content is assembled.

Associated with the topographical maps will be the series of thematic maps or overlays projected on the appropriate scale of topographical maps to present a particular theme. Some of the thematic information may already appear in rudimentary form in the topographical map, the extent being decided by the specification of the topographical map and the scope of the graphical presentation. The themes currently being mapped by official organizations are numerous, and it is logical to presume that many thematic maps are made because they depict data not sufficiently shown on traditional topographical maps. Any such series of thematic maps or overlays, for example land-use or forestry, will also be affected by any significant datum or geodetic network change, as also will the texts and lists in which the superseded coordinates, whether graticule or grid, have previously appeared. Recompilation and redrafting, a slow and expensive task by traditional methods, will be facilitated by automation.

Map libraries or geographic information centres contain the topographical and thematic maps so produced. These maps also find their way into the hands of the professional executive, administrator, planner, researcher, worker, and layman. The library staffs monitor the input and output of information, attempt to maintain the libraries up to date, undertake studies and research, and maintain reports to answer specific questions within their fields. Links are established with similar map libraries, both national and international, and with specialists and thematic libraries. In the language of the computer scientist, these libraries could rightly be called geographic, topographic or cartographic data bases.

The *positional accuracy* on the map of the depicted features will be dependent on the accuracies of the survey, of the compilation, and, in particular, of the manual cartographic drafting. The accuracies of manual plotting and drafting may be taken to be about 0.2 mm or 8/1000 in in the positioning of a point, and between 0.25 and 0.50 mm or 10/1000 and 20/1000 in in the positioning of a line.

These drafting accuracies affect the design of map symbols and the acceptable minimum distance between depicted features. They greatly influence the amount of information or map content which can be included in a map at any given scale, and hence the choice of map scale.

When the scale of a map is reduced, it becomes necessary to omit the less important features and to generalize the depicted information, either to make the map more easy to read or to separate or replace features which have merged with each other in the reduction of scale. For example, a road at the 1/1000 scale may be depicted by four lines to show the road edges and the sidewalks. At the 1/10,000 scale no distinction between the road edges and the sidewalks can be made, because a 2 m or 6½ ft-wide sidewalk, which measures 2 mm or 78/1000 in at the 1/1000 scale, measures 0.2 mm or about 8/1000 in at the 1/10,000 scale; and at the 1/50,000 scale a 30 ft-wide road would be shown by a single line of exaggerated width or by a double line in which the road width is even more exaggerated. With generalization, art has entered into the compilation and into the drafting of maps, and the role of the cartographer is increased.

Summarizing, the foregoing review of traditional topographical mapping practice mentioned some relevant points in the application of computerized automation:

> the fundamental place of the topographical map;
> the geodetic basis of the topographical map;
> the accuracy of the topographical map;
> the parent map scale;
> the great variation in the parent map scales between countries;
> the need to coordinate mapping at the parent scale between different levels of government;
> the family of map scales derived from the parent scale;
> the associated thematic maps;
> the generalization of the features at smaller scales;
> the changes of projection, grid, and scales;
> the changes of detail, control points, and datum, and their effects on graticule, grid, sheet lines, and map content;
> the concept of a map library as a data base, and of associated geographic, topographic, or cartographic data bases, and,
> the transmission of information between map libraries and between the various sources of information.

THE IMPACT OF COMPUTER TECHNOLOGY

INTRODUCTION

Near-earth satellites opened the way to the practical use of satellite geodesy for the fixing of ground control points on the earth's surface, and of satellite remote-sensing for the gathering of information. But of wider impact were the accompanying and the subsequent accelerated developments in electronic and computer technology, described more fully in earlier chapters of this book. From the beginning, it was apparent that topographical map production would be affected in the collection, storage, analysis, compilation and presentation of topographical data. Nevertheless, it has to be recognized that, by the 1960s, after more than 200 years of experience of national topographical mapping in some countries and of associated developments in the graphic arts, topographical maps had reached a high standard of accuracy and artistic presentation, which should be equalled or surpassed in any application of computer technology.

Moreover, many users of topographical maps and topographical map data were conscious of the opportunities provided by the developing computer technology of improving the production of derived and thematic maps, the making of descriptive diagrams, and the management of environmental data. They recognized that immediate benefits would accrue even working from the small-scale traditional maps to obtain geodetic and locational data of limited accuracy and to extract spatial data of a low resolution and of approximate locational accuracy. Just as the commercial banks and industries applied computer technology to the keeping of accounts and inventories using the simple identifications of name, number, and address or location, so organizations working in the environment such as property registries, utilities, traffic authorities, and census offices, adopted computer methods to record, maintain, and analyse inventories of items identified by name or number and by a rudimentary geographical location. Examples of these are discussed in subsequent chapters. For good reasons, these disciplines were more interested in the listing, accounting, and analysis of the data related to a feature, usually of a textual and numerical nature, than with a high positional accuracy of the complete feature. They were content, for the time being, with referencing a feature by, say, a single median point or by the two end points of a line, to which a list or inventory of the data of interest could be attached. Coarse cartographic diagrams or generalized maps at small scales sufficed to present the results. Experience was gained and practices were developed covering both graphic and computer procedures in the recording, maintenance, analysis, and presentation of environmental information. The effort and investment in these geo-coded activities and associated maps have increased progressively, and, in so doing, have made more urgent the need to provide the accurate topographical data in a computerized form and readily available to make base maps. The evolution in these disciplines will be towards a higher accuracy of geographical referencing and drafting, and to the recording of data at a finer resolution, for example, land-use per acre rather than land-use per 40 acres. Were these disciplines to move forward in advance of the geodetic and topographical surveys, they would be forced to establish for themselves the topographical base maps in digital form on which to overlay their themes. The authoritative topographical map situation would then become confused, and the subsequent duplicate production, correction and maintenance would be wasteful. There is already evidence elsewhere in this book that this is occurring.

The need for a common and accurate topographical base will become more necessary at the larger scales of maps, such as the parent scales of urban areas, where property boundaries will be demarcated by surveyed marks linked to the regional or municipal coordinate system. It is to be expected that the various professions, administrators, and workers will require that all computerized systems in their urban or built area will be compatible in the positioning of such data as property boundaries, buildings, utilities, etc.

THE DIGITAL MAP

Digital computing devices or computers operate on information expressed in numerical or digital form. A digital computer operates internally through the presence or absence of electrical signals like the ON or OFF of an electric current, and is therefore a two-state patterned device. Decimal digits and letters of the alphabet, i.e. alphanumeric characters, are expressed in terms of the binary digits '0' and '1', and binary codes have been standardized to facilitate the exchange of text between different makes of computers, e.g. the eight-bit American Standard Code for Information Interchange (ASCII), amongst others. However, numbers such as $x-y$ coordinates are readily expressed by a string of binary digits.

The computer has a capability of fulfilling a repertoire of instructions which is dependent on the manufacturer's arrangement of the computer's components and wiring. A computer operation is performed by entering 'instruction-words' which are converted into binary digits for interpretation by the computer; and by entering 'data-words' containing the data on which the instructions are to be performed, and which are also converted into binary digits. A selection of instruction-words and data-words to fulfil a purpose or to carry out an operation can then be written as a computer program. The results of the computer process are stored in the same binary digital form.

A *digital map* recorded in digital form can, consequently, be imagined, and the image may be more readily crystallized in the mind by stating the elementary steps in the production of a map by digital computer devices. These steps can be described as:

firstly, expressing the positions of the points and the lines on the map or source document as accurately located $x-y$ coordinates, e.g. of resolution 0.020 or 0.025 mm, and linking these coordinates to the graticule of longitude and latitude, or to a national grid, or to a local grid, or to all of these;

secondly, labelling or feature-coding these features in order to distinguish them;

thirdly, storing these data in binary digital form in some computer peripheral storage, e.g. magnetic tape, disk, or honeycombed cartridge storage; and

fourthly, retrieving the data to draft maps, to edit maps, and to provide textual data on drafting machines, visual display units, and line-printers, computer-routines being called as may be necessary for edit, symbolization, etc. Such retrieval may be direct or after the computer processing of the data, selecting, modifying, manipulating the data as may be required.

The digital data are stored in the digital data store or bank and are similar to the map data stored in the traditional map library. For every map, there is a file of digital data in the store or bank; and in a series of maps like a national or regional topographical series, the map series becomes a graphical index of the contents of the data bank.

THE ANTICIPATED EFFECTS ON TRADITIONAL MAPPING AND CARTOGRAPHY

Morrison and Rhind, in Chapters Two and Three respectively, have considered this in general terms but for topographic maps in particular the creation, collection, revision, analysis, presentation, and distribution of cartographic data, as well as map usage will be affected.

CREATION AND COLLECTION

The amount of cartographic and digital data is dependent on the specified map content, which is likely to be increased because of the greater accuracy and resolution of automated drafting, and the ease and speed with which digital data can be transmitted from the different collectors of information to the map-maker. There will be the capability of storing more information in the topographical digital data bank than in the traditional map, and more use can therefore be made of the data now found only in the thematic maps special to a discipline or profession, e.g. the classification of woods, forests, trees, and of land generally. Such information now published in thematic maps can, in the future, be held in thematic digital data banks, which can be associated with the topographical digital data banks. All the thematic information presented traditionally in thematic maps which are of interest to the topographic cartographer can be retrieved quickly. Descriptive textual information relating to the features can also be stored as digital data. Such a trend will also accentuate the importance of field completion surveys, will expose the weakness of relying on photogrammetric information and interpretation alone, and should influence the choice of the configuration of the digital mapping system.

REVISION

The speed and flexibility of the transmittal of digital data between data banks facilitate revision. Just as soon as the original cartographic digital data are revised in the parent scale or thematic data bank, the equivalent data can be corrected automatically in all the associated digital data banks, by the electronic transmission of the digital data, or, subsequently, from the digital data transported on magnetic tape reels. Consistency in the data stored in the different digital data banks is assured; and the procedure is quicker and more economical than the traditional practice.

ANALYSIS

The cartographic digital data are suitable for arithmetic and logical analysis by computer. The data can be transformed, reduced, interpolated, selected, and arranged in any manner which can be expressed or defined arithmetically, or decided by some logical tests. In cartography, this might include:

> changes of datum and projection;
> changes in the horizontal control points;
> changes of sheet lines;
> alteration of map scale;
> generalization of detail; the joining of map sheets and the partitioning of map sheets;
> the selection and extraction of individual features;
> the correlation of features in any mathematical formula or logical test;
> and the changes of the reference grid.

By such analysis, using the appropriate software, questions relating to the data can be answered.

PRESENTATION

The digital data can be presented in printed form as lists or textual data, or can be used to instruct drafting machines or television type displays to present maps or graphics. This has been more fully described by Boyle in Chapter Four. Electromechanical drafting machines of 1/1000 in accuracy are now available, for which input data of 1/1000 in resolution is to be expected, if the full accuracy of the drafting machine is to be utilized. But this is between two and four times better than the resolution of the average human eye at 12 in from the copy. Pinpointing is accurate to about 0.2 mm or 8/1000 in, which suggests that the locational accuracy of the digital data should be as good or better, if the maximum practical accuracy is going to be achieved. It is to be noted that if an accuracy of 0.2 mm is achieved, the map would be $2\frac{1}{2}$ times more accurate than the current specification for a Class A map at the 1/50,000 scale as Morrison points out in Chapter Two. Consequently, a 1/100,000 map drafted on a modern electromechanical drafting machine using suitable digital data would be more accurate than the 1/50,000 map produced by traditional manual drafting methods. As a by-product of the increased accuracy and higher resolution, features can be drafted more closely together, symbolization can be finer, and map generalization can be more delicate. Automation thus permits a review of the publication scale of mapping; of map symbolization; and of map content. The increased accuracy also makes feasible the miniaturization of film reproduction material for the purpose of reducing costs of reproduction and for use in automated navigation equipments.

Television type displays are likely to be used increasingly in special circumstances for a quick look at the graphical data for inspection, for design, for revision, and for generalization.

PERSONNEL

The introduction of computer technology will require a reorientation of the views of executives, and a reeducation of operators and craftsmen, which will be a continuing requirement in view of the rapid pace of development. Both Rhind and Morrison in earlier chapters have identified education as a critical variable. Electronic technologist support will be required for the in-house maintenance of the computer equipment and peripherals, usually termed the 'hardware'. In the further development or adjustment of computer programs, usually termed the 'software', the cartographer will require computer scientist support.

The application of computer technology to topographical map production is likely to:

> change the policies and programmes of topographical mapping;
> reorient the profession and craft of cartography in its practices and training;
> introduce electronic equipment to a degree which will require special maintenance personnel;
> demand a close relationship in production work between the mapmaker, the electronic technologist, and the computer scientist, which will be reflected in the organization table; and
> impinge on many disciplines, organizations, and activities connected with the environment. A great effort will be required to generate common goals, to achieve cooperation, and to avoid the proliferation of incompatible systems.

DATA AND PROCESSES OF AUTOMATED TOPOGRAPHICAL CARTOGRAPHY

INTRODUCTION

The data include:

> topographical data of the environment;
> topographical digital data collected at the digitizing stage from the source material;

cartographical data assembled by retrieving and transforming the topographical digital data of interest into the required projection and scale; and

graphic data formed by adding the symbols of the features and the relevant commands to instruct the drafting machine or display.

The processes include:

digitizing, editing, structuring, storage, retrieval, transformation, and presentation of data;

safeguarding of data for accuracy and against loss; and limiting of access to data to authorized persons only.

Cartographers, using the term in the widest sense, decide the operations to be performed in each process and select the data to be used. Computer equipment and the appropriate sets of instructions, i.e. computer programs, are required to enable the equipment to execute the operations. The requirements have aspects of cartography and aspects of computer science which need consideration. The application of computer technology is an art just as cartography is an art; consequently there are many solutions to the fulfilling of the requirements, and the best choice will vary with different circumstances of locale, and personnel, financial resources, and with the state of the technology at the time when the choice is made. Having selected the equipment and the configuration of the system, the task of writing the set of computer programs to instruct the equipment will be the responsibility of computer programmers.

AIMS AND REQUIREMENTS

The application of computer technology should fulfil the following aims and requirements in topographical map production:

collection and suitable storage of topographical digital data;
economic production of more accurate, more up to date, and higher quality maps;
recording of supplementary textual information about topographical features;
analysis and manipulation of topographical digital data;
retrieval of topographical digital data in whole or in part;
edit of topographical maps;
design of topographical maps and symbols;
drafting of topographical maps;
presentation of topographical maps on display screens; and
rapid transmission of digital data to other organizations.

MAP CONTENT

For each scale of topographical map, the features to be depicted and their manner of depiction or symbology will have been given in the map specification. The symbology varies between maps of different scales and between maps made by different map-makers, so giving scope to the exercise of cartographic art. However, standardization of the definitions and classifications of the topographical features depicted is necessary to improve the readability and the interpretation of maps, and to achieve compatibility between the different maps so that the topographical digital data can be coded or labelled uniquely for each feature. For example, not only should the types of roads, streets, and tracks in different regions be clearly depicted, but the method of classifying these types—motorways, first-class roads, second-class roads, unmetalled roads, etc., should be standardized over as wide a jurisdiction as possible.

The fundamental elements in the topography of the environment are the visible, natural, and man-made features, which can be regarded as entities at the scale of the map being produced. These entities have attributes. Cartographically, the topographical features are represented on the map by points, lines, and areas which are bounded by lines. The fundamental elements or entities can be combined to create other features. For comprehensive use by all disciplines, the fundamental elements should be stable and visible in the environment. Their combination will vary as the requirements of different disciplines vary and will be done to fulfil a particular specification of a topographical or thematic map. Thus, survey marks and boundaries can be combined to create land parcels.

The act of digitizing should, therefore, be regarded as the recording in digital form a topographical feature existing in the environment, and not merely as a point or line on a graphic for subsequent presentation. The stored data should be coded and arranged accordingly. At the 1/50,000 parent scale, this aim is not difficult to achieve for the fundamental elements at the 1/50,000 scale of depiction are readily selected and not numerous. Digitizing the centre line of a road and its intersections with other linear features, at the so-called 'nodes', in order to divide the road into segments between the nodes, will probably be sufficient for the definition of the road. Any changes in the character of the road, for example with or without sidewalks, change of width, or change of surface, may require to be recorded as attributes. This degree of definition is likely to fulfil all the requirements of users and thematic mapmakers interested in roads and working at the 1/50,000 scale.

At the larger 1/1250 parent scale occurring in urban areas, the problem becomes much more complex. Many professions are interested in the smaller details depicted at this scale—the man-holes in sidewalks and streets; the telegraph poles; the fire hydrants, etc. Again, the features should be recorded as entities which exist in the environment.

So many disciplines and professions have their special interests in the built or urban areas, and so great is the cost of collecting topographical information in the field by perambulation for the sole purpose of collecting such detailed information, that these disciplines and professions might well be left to collect their own specialized data and to record the data in digital form. It is assumed that a large scale topographical map of sufficient accuracy on which to base the specialized data will be available to them, preferably in digital form.

Compared with the more open and 'generalized' topographical maps at the 1/50,000 map scale and smaller scales, the digitizing of the larger scale maps of urban areas poses additional problems. At the map scales of 1/2500 and larger which are much used in civil engineering, the detailed nature of the depiction and the density of the features in urban areas make the task of digitizing all the features, along with the coding, both onerous and expensive. However, it is to be expected that local authorities, highway departments, and public utility, land development and architects' offices will be digitizing many of these features for their own purposes.

The opportunity is thereby given of limiting the topographical base map to a selection of features to act as a framework on which the features available from others in digital form can, subsequently, be overlaid. The digital data must be interchangeable, the accuracies of the digital data from the various sources must be in agreement at the scale of map compilation, and close cooperation among the producers will be necessary.

Implementation of such a system would greatly affect the current practice of producing the official large scale maps of urban areas. The map specifications, digitizing procedures, digital data structures, and method of cooperation will require to be developed and assessed. A suitable team drawn from the affected disciplines and professions and assembled in an appropriately equipped and staffed laboratory is indicated.

Under such a policy, the legal survey professions in those countries where property boundaries are defined by surveyed marks, might move to integrate the property plans in digital form with the large scale topographical maps produced of urban areas by computer.

DIGITIZING

For the purpose of this explanation, consider a table digitizer system connected through a small computer to the digital data storage or to another computer. The digitizer system consists of the table, the movable cursor, the coordinate display, the monitor display, and the operator's keyboard. A second visual display as marketed by some manufacturers might be included to display the work as it is being digitized. Similarly, a small digitizing table might be added for coding the features if the 'menu' method of coding, as described later, is used. The table digitizer surface can be regarded as being covered by an accurate grid of squares of about 0.24 mm or 1/1000 in side to provide a coordinate system for referencing positions on the table. The xy coordinates will be recorded at this resolution.

The orthogonality of the table coordinate system is periodically checked, usually at the beginning of each digitizing session by digitizing and so measuring the coordinates of a set of four points marked by the manufacturer and defining a large rectangle.

The source material, which may be a graphical compilation, a published map, or an air survey image, is placed on the digitizer table. Mounting the complete topographical compilation on the digitizer as the source material avoids the subsequent task of fitting the edges of portions of the map, but this is not always possible. The digital output from a photogrammetric plotter, for example, is restricted to each stereomodel, usually covering only a portion of the map. Additional procedures must then be taken to ensure that the edges of the stereomodels agree in the digital definition and in any graphical presentation.

The initial procedure requires the operator to follow a specified drill such as:

> sign on;
> command the procedure, which might be digitize or edit;
> enter map title;
> date; and
> operator's name.

The system is then ready to receive the first code of the procedure, which at the beginning of a new map will be the geographic referencing code, by which the xy coordinates of the table are related to the geographic coordinates of longitude and latitude, or the rectangular projection coordinates of the national, regional, or local grid, or all of these. The relationship is established by digitizing identifiable points on the source material, the ground coordinates of which are known. Two to four points will suffice for grid coordinates but more may be required for graticule coordinates where the graticule lines are curved on the source material.

Digitizing can then proceed. Each feature is feature-coded to identify its type, and its location is expressed in the xy coordinates of the table by pointing and tracking with the cursor.

Feature codes should be assembled in suitable groups to facilitate the subsequent retrieval and analysis of the digital data. Considering the case of a 1/50,000 topographical map production, assume that there are 400 or so different features, and that the following groups have been created:

> geographic reference data;
> relief;
> hydrography;
> vegetation;
> culture;
> communications—transportation;
> telecommunications;
> utilities;

boundaries;
aeronautical data;
map border data.

Within these groups, the features can be arranged. Under boundaries there might be: national, provincial, municipal, county, district, parish, park, electoral, etc.

Adopting a four-decimal digit number as a feature code, the first two might designate the group, and the remaining two-decimal digits would accommodate 99 features. A second set of four-decimal digits forming a minor code would permit the addition of attributes, such as the condition of a road, i.e. under construction, or closed, or the material of a bridge, i.e. stone, brick, steel, iron, or wood, or the height of a contour. Alphabetic letters may also be used in the codes to provide descriptions in formatted form or in unformatted form, which are stored with the xy coordinates of the feature.

To reduce keying errors in feature coding, and to accelerate the process of feature coding, various aids have been created. Using a 'function' keyboard permits the complete code to be entered on the depression of one key by the use of electrical plugboards and matching templates. Another type of aid is the use of a menu consisting of a set of square or locations drawn on a border of a digitizing table, or on a separate digitizing tablet. Digitizing within a location yields a pair of coordinates equated to the feature code.

However, the use of alphanumerics in the code allows a descriptive type of feature code to be used which can be easily remembered by the operator. The computer can be programmed to read such a code in its full or in an abbreviated form:

RELIEF or R　　　　　　VEGETATION or V
ELEVATION or E = 250　　etc.　　　　　(Linders, 1973)

The digitizing of the feature follows the entry of the code and attributes; one pair of xy coordinates for a point, and/two pairs of xy coordinates where an orientation is expected from the code, or for a straight line. A string of xy coordinates is required to define a winding line, and the speed of recording the points must be fast enough to keep pace with the speed of tracking and the output of coordinates. The following of line features on the surface of a graphic is neither quick nor accurate, so that many organizations digitize line features by digitizing discrete points at a frequency which the operator decides will define the line adequately. In the production of large scale maps where man-made features and hence straight lines and geometric figures predominate, the incidence of winding lines is not great and the slow procedure of digitizing discrete points is possibly economical. At the smaller scale of 1/50,000, where contours, hydrography, forest boundaries and natural features cause winding lines to predominate, the preparation of an etched plastic document from the source material enables the winding lines to be speedily and accurately digitized, and gives the operator a positive indication of the work completed and of any departure from the correct track by the scouring of the etched line. Fast tracking of winding lines, which may be up to 1 in per second, may require that as many as 50 points per second should be recorded to define the line.

The digitizing procedure is well summarized by a table of the data which might be collected and recorded.

> MAP SERIES, SCALE, SHEET NUMBER;
> DATE, TIME, OPERATOR NAME;
> GEOREFERENCE LONGITUDE LATITUDE XY COORDINATES;
>   EASTINGS NORTHINGS XY COORDINATES;
> GROUP;
> FEATURE CODE;
> ATTRIBUTE, TEXT;
> XY COORDINATES;
> END OF SEGMENT; END OF FEATURE; END OF MAP.

Statistical data such as the length of record, and the maximum and minimum values of xy coordinates in each record, can be automatically added to the record by the computer in the course of the processing, assuming that the computer capacity is available and that the appropriate sets of processing instructions, i.e. the computer programs, have been written and entered into the computer. Thus, it is possible to determined the total length of each type of straight and winding line feature for various purposes including the assessment of contract work. Other approaches using automatic digitizing by scanner are described by Boyle in Chapter Four.

PROCESSING

The data produced by the digitizing process are entered into the computer in the form of electric signals which can be interpreted by the computer. Within the computer data are organized in computer-words of a fixed number of binary digits, one computer-word being stored in each storage cell of the computer memory. The word length measured in binary digits is one factor in deciding the power of a computer. Other factors are the cycle time of operation, the speed of transmitting data internally, the size of core memory, and the quality and scope of the manufacturer's compiler program resident in the computer to convert the programmer's higher language written in alphanumerics into machine language written in binary digits at input, and vice versa at output.

The processing within the computer or computers may perform all the following operations:

> *Digitizing*   Prompt the operator in digitizing;
>   Link table coordinates to a georeference;
>   Rotate the recorded xy coordinates to common axes;

THE APPLICATION OF COMPUTER TECHNOLOGY

	Translate the recorded *xy* coordinates to a common origin;

Translate the recorded *xy* coordinates to a common origin;
Reject false feature codes;
Reject errors in sheet corner coordinates;
Reject points duplicated at digitizing;
Reject points too close together;
Send data to storage in an ordered manner;
Prevent the loss of data due to power failure.

Processing Arrange and index the data in storage to facilitate subsequent retrieval, edit, and expansion;
Add statistical data to the records;
Include links with associated features;
Edit the data by deletion, addition, and moving;
Manipulate the data for change of scale, projection, etc.;
Provide the required output to a drafting machine, display, or printer.

Data in storage should be arranged in a structured manner, which enables the data to be addressed with descriptive commands. Commands can be made to connect with a train of sequential commands in accordance with any selected hierarchy of data. The sequential commands can be displayed in succession on the monitor display to prompt the operator. Links or cross associations between the data in the hierarchy will transform the hierarchy into a network and permit greater analytical flexibility. The hierarchical structure of the data items in the Canadian Federal Surveys and Mapping Branch's automated cartographic system for the production of topographical maps can be represented diagrammatically. (Figure 5.1, Zarzycki, 1978).

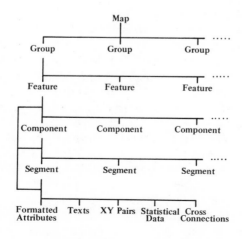

5.1 Hierarchical structure of data items

PRESENTATION

The stored data may be topographic data digitized from a photogrammetric plotter, from a cartographic compilation, or from a field survey. Alternatively, the stored data may be cartographic data digitized from a published map. The presentation of the data on a drafting table is dependent on the symbology given in the map specifications; and, in the case of a visual display unit, on the symbology selected by the operator. In the processing, a graphic file is produced from the topographic or cartographic file, and will include any modification of the data which may be required—data compression, interpolation, offsetting, scale change, or projection change. The computer programs select the appropriate symbols, provide the symbology, or choose the photographic drafting head, and control the drafting heads.

In an automated drafting system, there will usually be a data-point device, a computer controller, and an electromechanical drafting table, which will have a selection of drafting heads—single pen, scribing tool, multiple pen or scribing tool, and photohead. As input data, the system receives xy coordinates, either absolute or incremental, and the instructions for start, stop, speed, interpolation, choice of pen, choice of photohead aperture, and any other command needed to implement the manufacturer's software in the controller computer to operate the drafting table.

Drafting will proceed in accordance with keyed commands, or in accordance with the package of drafting commands and data in the graphic file prepared during the processing stage. The first drawing, probably in coloured ink, on paper, will be edited by the cartographer, and the relevant cartographic digital data on the original etched plastic document will be corrected at a digitizing station using the edit procedure. With or without further proofs, the final map can then be scribed or reproduction films made.

Maps drawn at a reduced scale are likely to need further edit to achieve the generalization required. The generalization of topographical features increases as the scale of the map decreases below 1/10,000. For example, the features at the 1/50,000 scale are greatly generalized, the topographical data surveyed at a larger scale being smoothed, moved, deleted, or replaced. In the traditional era, the modifications were done manually by a cartographer. In the automated era of cartography, some modifications such as smoothing deletions can be done, in whole or in part, by computer-processing; but other modifications will only be possible after a visual examination of the graphic presentation. The editor's corrections and generalizations can be viewed as a whole on a proof of the drafted map, which will present the intended appearance of the published map. These corrections and generalizations, being depicted graphically at the scale of the drawing, are conveniently digitized on a table digitizer. In so doing, the digitizing operator will bear in mind the effect of the corrections and generalizations on the photomechanical stage in the reproduction of the map, which includes the provision of tint fillings to some features and the making of separate printing plates for each colour.

A repeatability of 0.25 mm or 1/1000 in and an absolute accuracy approaching this figure are being achieved in the best electromechanical tables available. Drafting speeds reaching 25 cm or 10 in per second are also being achieved.

The provision of data for a cathode ray tube type of visual display unit of about 0.15 mm or 6/1000 in resolution is similar to that for a drafting table. If edit of the display image is to be undertaken, then the appropriate visual display unit must be directly connected to a computer with adequate edit programs to perform 'interactive' graphics. Visual displays provide a quick look at the graphic data, and are suitable for inspection, for design, for revision, and for some generalization of the graphic image. In such displays, the image can be sectionalized, i.e. 'windowed', and a higher resolution achieved by enlarging the section or window of interest to the full format size of the visual display.

SYSTEM CONFIGURATION, EQUIPMENTS AND PERFORMANCES

INTRODUCTION

Systems may be off-line or on-line, or may be partly off-line and partly on-line. The configuration will be influenced by the developments in computers and in computerized equipment as described by Boyle in Chapter Four, and by any special requirements of the production of maps. Current table digitizers and electromechanical drafting tables are accurate and reliable. As Boyle has indicated, future development will lower costs, accelerate digitizing, and add microprocessors to undertake standardized processes, which will enable a smaller main computer or selection of minicomputers to undertake the remainder of the processing.

Organizations carrying out large programmes of topographical mapping by photogrammetric compilation may modify some stereoplotters to give their output in digital form. The Canadian Federal Surveys and Mapping Branch has a system configuration which includes both photogrammetric digitizing and table digitizers, and the evolving system will permit both off-line and on-line operation (Zarzycki, 1978).

Alternative approaches to digitizing, taken to accelerate the action of digitizing, are raster-scanning with the subsequent vectorizing and feature-coding of the data, and, secondly, automatic-line following by electronic means, although Boyle argues that this latter approach is less promising.

The topographical digital data collected by any of these means are of a high accuracy and resolution. Selected items and portions of the data should be retrievable for use in the making of other maps and map overlays.

CONFIGURATION AND EQUIPMENTS

The three stages of digitizing, processing, and presentation can be undertaken as independent operations on separated unconnected equipment systems, i.e. 'off-line', as when the digital data are recorded on magnetic tape reels at the output of each stage, and the magnetic tapes are then used to provide the input to the next stage.

The off-line mode reduces the initial capital investment, and allows a gradual approach to the adoption of automated techniques. But the mode does not allow a real-time or rapid revision and correction procedure to operate at all stages of the production line, and increases the total production time.

The series of operations can also be conducted 'on-line', the equipment of the digitizing, computer processing, and drafting or presentation stages being interconnected by data channels. The on-line mode permits the real-time processing, edit, and drafting of the digitized data, makes possible an immediate choice of output from the stored digital data and provides more scope for introducing operator prompts, error checks, and access controls to safeguard the data. However, the on-line system requires additional computer capacity and computer programming within the production line, which means greater capital investment.

Off-line systems need not necessarily be small. In the Ordnance Survey of Great Britain, the separated stages of digitizing, processing, and drafting in 1978 were:

Digitizing	nine table digitizers with digital data output to their respective incremental magnetic tape units; and
	five table digitizers with digital data output to one magnetic tape unit through a PDP-11/05 minicomputer.
Processing	the organization's main computer—ICL 1906S.
Drafting	two Xynetics high-speed flatbed plotters, each controlled by a Hewlett Packard 2100 minicomputer; and
	two Ferranti Master Plotters, each with a light spot projector and controlled by a minicomputer.

Large-scale topographical maps at 1/1250 scale are produced. From the resulting digital data, 1/10,000 scale and 1/25,000 scale maps have been derived, contours being added. Generalization by computer at these smaller scales is limited to the omission of minor features. Other generalizations and corrections of blemishes are undertaken manually before printing the final product (Bell, 1978).

On-line systems of a small size are marketed by some manufacturers, for example, Kongsberg of Norway, Ferranti of the United Kingdom, and Gerber of the United States, amongst others. These systems, controlled by a 16-bit word minicomputer, may have one or more digitizing stations, a plotter, a disk unit, a tape unit, and a display. The amount of processing is restricted by the size of the computer, and digitizing and drafting are usually done at different times.

A large on-line system has been developed for topographical cartography called the 'automated cartographic system', in the Canadian Federal Surveys and Mapping Branch. The system consists of:

Digitizing — five table digitizers, three of 0.025 mm resolution and two of 0.020 mm resolution, and up to sixteen more can be added; the digital data output is collected by a PDP-11/45 (32K memory) minicomputer, standardized at either 0.025 mm or 0.020 mm resolution as desired, and then transmitted to the main computer, or sent to storage on magnetic tape in the case of failure of the main computer;
similarly, digital data received on-line or off-line from remote stations such as the Geological Survey of Canada, are received by a PDP-11/20 minicomputer and passed to the main computer.

Processing — a DEC 10 computer with 192K memory, random access disk storage of 100 million bytes, and magnetic tape drives.

Drafting — two high-quality electromechanical drafting tables, one Kongsberg Kingmatic 1215, and one 1216, which are controlled by a PDP-11/45 (32K memory) computer; the digital data received by the PDP-11/45 computer from the DEC 10 main computer, are then sent to the drafting tables.

(Zarzycki, Harris, and Linders, 1975)

A number of computer programs reside in the main computer for file management, digitizing, editing, and drafting.

The *file manager* program manages the data; guards access to the data; retrieves, stores, and sorts the data; maintains the structure or hierarchy of the data; records titles, dates, times, and operator names; supports a powerful command language; and reads data from magnetic tape and writes data on magnetic tape.

The *cartographic monitor* program relates the xy digital data to a geographical reference, processes the digital data in the manner required, and provides the edit functions.

The *graphic* program creates the display file of instructions, data, and symbology for the drafting machines.

These programs occupy about 37K of computer memory. Allowing 35K of memory for the manufacturer's operating system, the programs require 72K words (36 bit) of computer storage. Additional storage is needed for the data and its processing. A total of 192K words of computer storage has been provided.

No attempt will be made here to choose between off-line and on-line systems, because the circumstances of each requirement vary so much, the technology is moving forward rapidly, as Boyle indicates in Chapter 4, and the relative costs of hardware and of software, are changing fast, the former decreasing and the latter increasing. An approach will be made to the selection of a system for general topographical cartography and this will be described. Before doing so, however, the influence of map compilation by photogrammetric methods needs to be considered.

PHOTOGRAMMETRIC COMPILATION

Many organisations and commercial firms are mainly engaged in compiling new maps from air photography using photogrammetric stereoplotters. The graphic compilation produced in this way can be reproduced as an etched image on a stable plastic sheet for digitizing on a table digitizer, as previously described. In this procedure, the features of the map are tracked twice: first, on the stereoplotter and, secondly, on the table digitizer.

Modification of the stereoplotter to give a digitial output in xy coordinates of the position of the floating or measuring mark would eliminate one tracking operation, and any loss of accuracy in drafting the graphic compilation on the stereoplotter drawing surface or pantograph. Many manufacturers now sell stereoplotters equipped to give the xyz positions of the floating mark, and to display the tracking on a cathode-ray-tube display screen.

In Toronto, the Ministry of Transportation and Communications of the Ontario Government produces 1/500 and 1/2000 scale engineering plans of transportation corridors and of the adjacent topographical details of interest. The features are digitized directly in four Zeiss Planimat stereoplotters, lines being defined by a string of discrete points. The need for a high accuracy of position and the predominance of geometric features in the large-scale plans, influenced the decision to adopt the practice of plotting most detail by discrete points. Contours and boundaries of woods are digitized in time interval mode. The digitized xyz data are also used directly in highway design calculations.

Corrections to the digitized stereocompilation are made by feeding the data of feature code, xy coordinates, and text as recorded on magnetic tape, to a high-resolution display system—the HRD1 Laser Display produced by Laser Scan Ltd, of Cambridge, England. The HRD1 has a very large storage display having a screen measuring 1.0×0.7 m, and a limited capability of displaying a non-stored or 'refresh' image, e.g. the image of the cursor movement during the revision of a map. It is driven in this instance by a PDP-11/45 minicomputer. The manufacturer provides the software for interactive editing and edge-fitting. The data are recorded graphically at a reduced scale by a laser beam spot of 0.02 mm or 0.008 in on a photochromic film. After ten times enlargement, this image is projected on the display screen or, alternately, a permanent copy can be plotted on micro-fiche. To achieve a higher accuracy in edit operations, any smaller rectangular area of the photochromic film image can be 'windowed' and enlarged by the computer to the full photochromic image size, which will fill the display screen on projection. Corrections made by the operator by the movement of the cursor on the display screen should be equivalent to an accuracy of 0.1 mm or 4/1000 in on the photochromic film. The corrected data are used to draft the final large scale plan on a high-quality Gerber 1232 plotter using a photohead (Turner, 1978).

The federal topographical mapping activity in Canada is directed, primarily, to completing the 1/50,000 map coverage and to revising the maps already published, and is executed by photogrammetric methods. The advantages offered by digital mapping have aroused interest in the Canadian Federal Surveys and Mapping Branch in collecting digital data at the stereoplotter, as an alternative to collecting the data by digitizing the graphical stereocompilations on table digitizers. A photogrammetric digitizing system was sought which, as far as possible, permitted the photogrammetric operator to provide correct data, to remove blemishes, and to ensure smooth joins between segments of lines and at stereomodel edges. Three Wild B8 stereoplotters have been modified to give digital data output. These along with a table digitizer have been connected, for the purpose of collecting, storing, manipulating, and editing the digital data, to an inter-active graphics design system (IGDS) manufactured for the market by M & S Computing Inc., Huntsville, Alabama, U.S.A. The IGDS equipment includes two video graphical display units, a PDP 11/45 or a PDP 11/70 computer and an RPO4 disk arranged for the sequential storage of about 80 megabytes of digital data and for data retrieval by the manufacturer's (M & S) disk scanner. It is anticipated that the IGDS can support up to six Wild B8 stereoplotters as well as a table digitizer. Concurrently, a Gestalt Photomapper GPM 2/3 has been installed to provide a close grid of ground elevations from which contours at 10 m interval for 1/50,000 scale maps can be generated by computer. The method is suitable in areas of Canada where the terrain is free of vegetation. Associated with the Gestalt Photomapper are two Wild B8 stereoplotters, enhanced as already described and a table digitizer system to provide the hydrographic features and other planimetric information (Zarzycki, 1978).

Both these systems are currently operated off-line and produce digital data recorded on magnetic tape or on disks in the manufacturer's data format. After conversion, the data can be fed into the automated cartographic system in the manner of data from a remote station as previously described.

PERFORMANCE

Figures of performance vary. They are dependent on the ability and training of the operators, on the quality of the individual equipment, on the procedures adopted, and on the density and type of map content. The following figures of the production of 1/50,000 maps of open hilly country are given merely to illustrate a method of comparing performance by different procedures.

Annual output of 1/50,000 maps working two shifts per day
    One stereoplotter and two operators per day   16 maps
    One digitizing stereoplotter and two operators per day   12 maps
    One table digitizer using etched source material and two operators per day   50 maps
    One electromechanical drafting machine drafting an average of $1\frac{1}{2}$ proofs before producing the map, and two operators per day   300 maps

Calculating from these figures, the production of 272 maps in a year would require:

    by table digitizing on etched source material: seventeen stereoplotters and six table digitizers; and
    by digitizing stereoplotters: twenty-three digitizing stereoplotters.

The number of operators is the same in each case.

In determining the cost/benefit factor, the cost of purchasing the six additional stereoplotters and enhancing twenty-three stereoplotters should be equated to the cost of the system of six table digitizers and to any benefits, which the digitizing stereoplotters provide, in particular the increased accuracy.

It is apparent from such a calculation that the direction of development should be to simplify and, thereby, to accelerate the operation of a digitizing stereoplotter so that the addition of the six stereoplotters can be avoided. The resulting saving of the cost of twelve operators can then be applied in the above equation to help to compensate for the cost of enhancing seventeen stereoplotters.

ACCURACIES

The traditional production of 1/50,000 scale maps is aimed to achieve a locational accuracy of 0.5 mm or 20/1000 in in the published map. In such production, using say a Wild B8 stereoplotter, the accuracies of the various procedures approximate to the following figures:

    in the graphical stereocompilation of lines: 0.2 mm or 8/1000 in at the 1/50,000 scale;
    in the manual drafting of lines: better than 0.5 mm or 20/1000 in at the 1/50,000 scale.

But 1/50,000 production from digitized data could achieve the following accuracies:

digitizing an etched stereocompilation and machine drafting: 0.20 mm or 8/1000 in at the 1/50,000 scale;

digitizing on a digitizing stereoplotter and machine drafting: 0.100 mm or 4/1000 in at the 1/50,000 scale.

In both methods the digital coordinates of the positions of the manually moved stereoplotter's tracing device are obtained. In the first method, a graphical compilation is drawn either directly on the plotting table of the stereoplotter or after a scale reduction using a pantograph, and the graphical compilation is subsequently reproduced as an etched document and digitized on a table digitizer. In the second method, a digital output of coordinates giving the positions of the tracing device is obtained directly from the digitizing stereoplotter. Some loss of accuracy due to the extra stages is to be expected in obtaining the digital data from the graphic stereocompilation, but with suitable procedures the loss can be minimized. The accuracies can be compared by the successive digitizing of a well-defined line in the stereo-image by each method in a practical experiment.

The cartographer has the option of replacing the 1/50,000 scale map with a 1/100,000 scale map which would provide information of the same locational accuracy, or of using the same data to publish larger-scale maps, any additional map content appropriate to the larger scale being added by revision methods. 1/20,000 scale maps of the required accuracy could be produced from the digitized data and, in the case of the digital data of even higher accuracy produced by a digitizing stereoplotter, 1/10,000 scale maps could be produced. Moreover, the specified national standard of accuracy of a 1/10,000 map may be lower—0.825 mm or 33/1000 in is quoted for map scales larger than 1/20,000 in the American Society of Photogrammetry's *Manual of Photogrammetry*, third edition, vol. II.

The achievement of these higher accuracies depends greatly on good operation of the equipment. The edit, correction or revision of data may consist of deleting xy coordinates of 0.025 mm or 1/1000 in resolution, substituting xy coordinates of the same resolution, making smooth joins between lines, and ensuring that each feature is uniquely defined by the digital data recorded. There must not be two different sets of digital data to define the same feature. Interactive methods on a table digitizer or on a visual display screen give the operator great freedom to change the digital data as well as the graphic. Using a visual display screen of 0.15 mm or 6/1000 in image resolution, the relevant portion of the map is isolated or 'windowed' and enlarged six times, say, to achieve the required accuracy. Within a computer-programmed tolerance of locating a feature, features or segments of lines are located by the appropriate coding and digitizing of a point on or near the feature; that is, 'proximity pointing'. The feature is then deleted; but to guard against the loss or the degradation of the best data, due to an erroneous command or to poor digitizing, a copy of the deleted data should be made automatically in the computer and stored until the next revision, unless its deletion is approved by a specified supervisor.

OVERLAID THEMATIC MAPS

The higher accuracy of digital data affects the production of thematic maps based on topographical maps produced by digital methods. Identical features of the topographical map and thematic map should not be digitized twice for reasons of economy, accuracy, speed of production, and uniqueness of digital definition.

Suppose that a land-use map or overlay is to be made on the foundation of a 1/50,000 topographical map, and that the roads, tracks, railways, rivers, streams, coastlines, and such line features depicted on the topographical map, form the boundaries of the land-use areas and have been recorded in the computer data storage in binary form as xy coordinates, related to the national grid. In the production of the thematic product, keying the appropriate commands and digitizing, within the programmed tolerance, the boundary intersections, which can be termed 'nodes', suffice to locate in data storage and then retrieve the topographical xy coordinates defining each boundary of the area of interest for subsequent processing. The process is similar to extracting the superseded data in the revision procedure already described. The nodes are points in the topology of the environment, and extend the scope of organizing and retrieving topographical digital data in storage. The topographical and the thematic digital data should be expressed in the same or related coordinate systems and be of comparable accuracy.

APPLYING COMPUTER TECHNOLOGY TO TOPOGRAPHICAL CARTOGRAPHY—AN APPROACH TO THE IMPLEMENTATION OF A SYSTEM

INTRODUCTION

The organization contemplating the introduction of computer technology to topographical cartography may be governmental, national, provincial, or municipal; or commercial. It may be assumed to be experienced in the traditional methods of topographical map production and will, consequently, be faced with the task of digitizing new maps and maps already published. It is also assumed, in this first approach, to have a large enough production commitment, sufficient financial resources, and suitable staff and conditions to support a minimum on-line automated cartographic system to undertake all its production.

Successful automated cartography must, by definition, be successful not only in its cartography but also in its automation, and must fulfil the relevant production aims of the organization economically and within the specified time. It is logical to suppose, then, that a project team might be formed within the organization to determine the requirements and to plan the implementation of the automated cartographic system; and that the team might consist of:

- *a professional cartographer*, experienced in the cartographic production, practices, and procedures of the organization;
- *a computer scientist*, knowledgeable in the current and developing state of computer technology, capable of advising on the hardware and software needs of computer graphics, and able to direct a group of computer programmers in the subsequent writing and maintenance of the cartographic computer programs, which will be a continuing task in any organization wishing to be flexible in its production; and,
- *a managerial professional executive*, closely in touch with the organisation's management, versed in the cartographic needs and in the possibilities of automation, able to propose practical objectives, and to manage the project to achieve the aims within the limits of time and finance set by management.

PURPOSE

Let it be presumed that the purpose is:

- to accelerate the production and revision of maps in order to achieve more up-to-date maps;
- to increase the accuracy and quality of maps;
- to provide digital cartographic data;
- to store more information of the topography; and,
- to reduce the cost of cartographic production.

Consideration will be given to the three stages of data collection, data processing and storage, and data presentation.

DATA COLLECTION

Being a general map production organization obliged to collect digital data from published maps as well as from new maps, a data collection system of table digitizers connected on-line to a main computer, offers proved equipment with the necessary flexibility at a minimum capital cost, and will suffice to demonstrate this approach. The addition of other more complicated and more expensive digitizing systems, e.g. digitizing stereoplotters or scanners, can be considered separately and subsequently as these alternatives become technically simpler and cheaper.

It has already been stated that one modern electromechanical drafting table can produce 300 map sheets of 1/50,000 scale and 30 minutes by 15 minutes format of open hilly country in one year working two shifts each day. To achieve a balanced production line, digital data must be collected at no less a rate. Again, it has been stated that an on-line system of six table digitizers has the capacity to digitize 300 map sheets of this type in 1 year working two shifts each day. The six digitizing stations each consist of a table digitizer, a movable cursor, a coordinate display, and an operator's keyboard with a visual display monitor. An interactive visual display unit might, with advantage, be added to one of the digitizing stations to provide an additional facility for design and generalization. The other stations could be similarly enhanced if the type of production so demands.

In such a data collection system connected on-line, the digital data from the table digitizers are suitably collected by a 16-bit-word minicomputer of 32K memory, which may check orthogonality of the table, validate the feature codes, reject duplicated coordinates, collect points at the required minimum separation, translate and rotate coordinates, enable some digitizing corrections to be made, and supply some operator prompts. The collected data are then passed to the main computer for storage in an ordered manner and for further processing. An alternative output to a magnetic tape unit guards against the loss of digital data due to any failure of the main computer. The rapid development of more powerful microprocessors, and the decreasing costs of minicomputers and of medium-size computers, will influence the configuration and the choice of computers, which the computer scientist in the project team will need to appraise. The configuration given in Figure 5.2 is based on the performance of computers in recent years and on current costs.

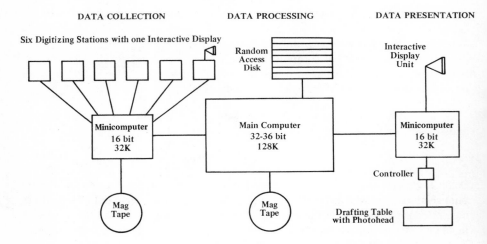

5.2 On-line computer-assisted cartographic system (after Zarzycki 1978)

Current costs of such a data collection stage consisting of six digitizing stations with one interactive display unit, a 16-bit-word minicomputer of 32K memory, magnetic tape unit and connecting data channels should not exceed $200,000.

DATA PROCESSING AND STORAGE

Part of the processing is undertaken in the minicomputer at the data collection stage, just as some of the processing is done in a minicomputer at the presentation stage.

The bulk of the processing, however, is done in the main computer, and will be concerned with the data input and output, the data management, the creation of cartographic files for the graphical output with the associated analysis, and edit. Cartographic digital data management has its own characteristics, and the 'file manager' should be designed accordingly to achieve flexible, rapid, and economical data management. Although features may be digitized in a haphazard manner, they should be arranged logically in a hierarchical index or directory, as previously described, and the data stored in an organized manner on disks, like the index and shelves of a book library. Having ascertained the complete cartographic requirement, it will be the computer scientist's task to evaluate the type and size of computer system—the main computer and the peripheral computers including the apportioning of the processing between them. In this approach, a main computer of 128K memory with a word length of not less than 32 bits, and a 300 megabyte disk system are proposed. Permanent digital data storage will be on magnetic tape. The hardware cost of such a system is likely to be about $300,000.

DATA PRESENTATION

The equipment includes the automatic drafting machine and its controller, and an interactive display unit connected on-line to the main computer through a 16-bit-word 32K memory peripheral minicomputer. The minicomputer is suitable to undertake the following processing, so decreasing the load on the main computer:

> Partitioning or 'windowing' the map;
> Choosing a positive or negative image;
> Choosing the size of symbology;
> Interpolation, compression, and offsetting of data;
> Selecting the drafting tool and photohead aperture and orientation;
> Scaling of x and y dimensions;
> Specifying the origin of the drafting table;
> Emplacing the drawing on the drafting table in position and orientation;
> Specifying the speed of drafting.

These instructions are converted by the controller of the drafting table into drafting table operations and movements. The cost of the equipment in such a presentation stage approximates to $180,000.

STAFF

The staff required for the automated cartographic production of 300 map sheets per year of the type described, working two shifts each day, is estimated to be:

Data collection	2 supervisors
	12 operators
Data processing	2 computer operators
Data presentation	2 supervisors
	2 operators
Computer support	1 computer scientist
	3 computer programmers
	1 electronic technologist
Total	25

For comparison, the staff required to produce the same work by traditional methods:

Drafting	4 supervisors
	38 drawing personnel
Total	42

The resulting reduction in personnel is 17, but due to the varying salary rates between the computer staff and the drafting staff a more realistic figure to use in the comparison is 16.

PROGRAMMING

The computer programming staff will be engaged in the maintenance and in the enhancement of the software. To produce the initial cartographic software for data collection, file manager, edit, and data presentation, designed specifically for the cartographic production requirement, a capital sum of up to $200,000 is suggested for contracting the work to a university or to industry under the direction of the project team's computer scientist.

COST BENEFIT SUMMARY

Hardware at $680,000 and software at $200,000 give a total of:	$880,000
which amortized over 7 years produces an annual repayment of:	$125,714
Averaged annual interest on loan at 12%:	$52,800
Equipment maintenance by manufacturer at 7½% per year and some software enhancement by contract work:	$66,000
Total	$244,514
Allowing for the permanent magnetic tape storage, which will increase progressively, let the annual cost be taken to be:	$260,000

The effective reduction of 16 personnel at an average annual salary of $A, and an overhead factor of 2.2 to cover

accommodation, pension, and other fringe benefits, yields an
annual saving of: $35.2A

In order to achieve a saving on conversion to automation, the
average salary of the drafting staff must not be less than: $7,386

The above calculation depends on the greater speed of digitizing an etched plastic source document compared with the speed of manual scribing. Operators vary in the speed of digitizing just as they vary in the speed of manual scribing. A figure of 25 map sheets per operator per year has been used in this calculation, but experience has shown that this figure might be increased by 40% by a fast digitizing operator. Five digitizing stations would then be more than enough to undertake this programme of work.

Speedy and accurate digitizing made possible by the use of an etched source document becomes increasingly more advantageous compared with digitizing by discrete pointings as the map content's proportion of straight lines decreases and proportion of sinuous lines increases. In this illustration, it is to be expected that the professional cartographer in the project team will conduct pilot projects to determine digitizing times and to develop selection procedures and training schemes for digitizing operators.

Apart from the speedier and more economical production, the other benefits of applying computer technology to cartography which should be equated to the costs are:

- the provision of digital topographical data for other users;
- the ability to select the format, scale, content, and projection quickly from the source digital data to make derived maps;
- the convenience and speed of edit so facilitating the achievement of more up-to-date maps;
- the more rapid transmission of data, and hence
- the easier maintenance of consistent data in all libraries, data bases, and geographical centres.

THE CARTOGRAPHIC DATA BASE AND OTHER TRENDS

THE DATA BASE

The procedures of data storage, access, retrieval, sorting, association, and modification required for the adequate management of digital data in storage have attracted much of the recent effort applied to improving software for commercial applications, and for some environmental purposes. The creation of a topographical digital data base will become increasingly important in applying computer technology to topographical cartographic processes as users become more interested in digital cartographic data. As maps and source material are digitized for production, the digital data should, therefore, be defined and structured in a manner which is suitable for the subsequent development of a digital data base. In a digital data base, the data are stored as entities without any need for the user to know their physical location in storage. The user merely specifies the data elements and the particular associations between these elements which are of interest. The data management software then accesses the data, brings the elements into association, and fulfils the user's requirement or provides the answer to a query.

The essence of the design of the Canadian Federal Surveys and Mapping Branch's 'automated cartography system' is its usage for map production of all kinds and its suitability for subsequent development into a digital data base, for which the topographical map will be an index. The essentials are the treatment and coding of topographical features as entities existing in the environment; their arrangement in a hierarchy—map, group, feature, component, segment, etc., which also permits a hierarchical sequence of commands and subcommands, where necessary, to be created, interpreted, and displayed by the computer, using a dictionary of these commands, to guide the operator. An example of such a sequence of commands displayed successively on the operator's monitor display screen would be:

> bridge; road, rail, or foot;
> over road, over rail, over river, over cutting, over buildings;
> steel, stone, wood; etc.

Further essentials are:

> a suitable geographical reference system for locating and interrelating geographical data;
> the recording of features to a known locational accuracy;
> the adding of descriptive texts to features, both formatted and free prose;
> the computer capacity and data management programs to extract, rearrange, manipulate, store, and present data in the manner and at the time needed.

At the medium topographical scale of 1/50,000, a pattern of nodes and segments (also termed spokes or chains) can be readily visualized in rural areas, where this scale may be the parent scale. Various listings or directories of the groups, features, and segments in the digital data storage can be imagined. The nodes can also be recorded, numbered, extracted, and listed separately in any methodical order for processing purposes. Segments, spokes, or chains, can be combined to form a feature, and can also be joined to create elemental areas like land parcels, and composite areas like golf courses. The structure adopted, and the file management software developed, will depend on the range of the uses of the digital data base. The task becomes more complicated as the map scales become larger and the map content more detailed, if a comprehensive digital data base of high locational accuracy is sought. At this time, governmental topographic mapping organizations are paying prior attention to the needs of automated map-production. Until topographical map data become generally available in digital form over wide areas, the creation of a digital topographic and cartographic data base will not be urgent.

OTHER TRENDS

Improvements of computer equipment will continue, microprocessors are becoming more powerful and more flexible and will replace some of the processing now undertaken by minicomputers; and computer equipment will become cheaper. The conversion of stereoplotters to provide digital output should, as a consequence, become cheaper.

Automatic digitizing equipment by laser beam and fast drafting equipment by laser spot are currently marketed by Laser Scan Ltd., of Cambridge, England, and attract interest. In their FASTRAK digitizer, a reduced negative typically one-fifth of the size of the original material to give a minimum line-width of 30 μm is interrogated by a laser beam probe. Each selected line feature is scanned 500 times per second at intervals of 10 to 15 μm. The scanning laser beam probe travels along a line at the equivalent of $1-1\frac{1}{2}$ in per second at the scale of the original source material. The software is capable of dealing with line intersections, but in difficult cases operator intervention is necessary. The firm of IOM-TOWILL of Santa Clara, California, U.S.A., provides a digitizing service using laser equipment.

The vectorizing of raster-scanned data may become a competitive alternative in the provision of digital data, but the separate step of feature coding will still be required.

The overall effects of these developments are to increase the accuracy of the digital data, and to accelerate the processes of digitizing and drafting.

Software quality and improvements have not been so satisfactory, and there is scope for refining software to provide more flexibility, to reduce the computer storage used, to accelerate the processing, and to facilitate the interchange of data between systems.

Once the production of topographical maps by national organizations is automated to the extent of applying computer technology to the production processes, the current national standards of map accuracies, and the choice of map scales in the national family of maps should be reviewed. Surveyed data previously acceptable for a particular scale of map may no longer be of sufficient accuracy. User requirements previously satisfied by a specified map scale may then be satisfied by a smaller map scale, requiring fewer map sheets for the area of coverage. The parent or basic scale of topographical survey may then be better expressed by the accuracy of the digital data locating the topographical features rather than by the parent or basic scale of topographical map depicting the topographical features.

In such a fast moving and vibrant technology, close cooperation between the cartographer and the computer scientist must be maintained, in order to ensure, on the one hand, that as far as possible the best, latest, and most economical aids are employed using computer technology, and, on the other hand, that good cartography is not stunted by too impulsive an application of computer processes without a proper appreciation of the characteristics of good cartography as a means of communication.

REFERENCES

Bell, J. F. (1978). 'The development of the Ordnance Survey 1/25,000 scale derived digital map', *The Cartographic Journal of the British Cartographic Society*, **15**(4), 7–13.

Linders, J. G. (1973). 'Computer technology in cartography', *International Yearbook of Cartography*, **XIII**, 69–80.

Turner, J. G. (1978). Personal Communication. (Ontario Government, Ministry of Transportation and Communications, Surveys and Plans Office.)

Zarzycki, J. M. (1978). 'Digital mapping and computer-assisted cartography', Conference of South African Surveyors—CONSAS, Capetown.

Zarzycki, J. M., Harris, L. J., and Linders, J. G. (1975). 'Topographic-cartographic data base and automated cartography in Canada', Conference of Commonwealth Survey Officers.

The Computer in Contemporary Cartography
Edited by D. R. F. Taylor
© 1980 John Wiley & Sons Ltd

Chapter 6

Computer-assisted Cartography: Research and Applications in Sweden

LARS OTTOSON AND BENGT RYSTEDT

INTRODUCTION

There are three basic requirements for computer-assisted cartography (C.A.C.): coordinate based data, methods for handling this data and building a numerical model of the map, and technical equipment to allow both input and output. Sweden was a pioneer in research in C.A.C. with work on all three of these elements beginning in the mid-1950s.

The first step in computerizing topographic mapping in Sweden was to digitize terrain data and then redraw it using an automatic drafting machine. At an early stage, however, a decision was taken to wait for the development of efficient methods of digitizing large quantities of data and to introduce computerization step by step. The first part of this chapter describes research and development on computer-assisted topographic mapping in Sweden and complements the discussion by Harris in Chapter 5.

The aim of thematic mapping is to illustrate geographical aspects of particular themes or topics, often with a topographic map as background. Since in Sweden there are a great number of registers containing information which can be related to persons and real properties it is possible to tie statistical information to geographical areas at various levels and scales such as counties, municipalities, parishes, and real properties. By recording coordinates for these areal units, flexible computer-based methods can be used to obtain geocoded information.

Research on pure cartographic aspects of computer-aided thematic mapping has not been extensive. More attention has been paid to the creation of coordinate based data and the construction of numerical map models. Part Two of this chapter deals with these efforts in Sweden.

PART ONE: COMPUTER-ASSISTED TOPOGRAPHIC MAPPING

INITIAL REMARKS

In Sweden the responsibility for topographic mapping is divided between the State and the communities in such a way that the State is responsible for map production at the scale of 1/10,000 and smaller, and the communities for maps at larger scales. Thus the National Land Survey of Sweden produces a number of official maps at scales smaller than 1/10,000. The main series of this production is the economic map (a general land-use map) at 1/10,000 or 1/20,000, the topographic map at 1/50,000 or 1/100,000 as well as general maps at the scales of 1/250,000, 1/500,000 and 1/1,000,000.

Computer-aided techniques were introduced in Swedish topographic map production in the middle of the 1960s (Ottosson, 1978a). At the same time, the first flatbed plotter was installed at the National Road Board. This plotter was used for development work and tests by other authorities interested in computerizing different cartographic activities. At the end of the 1960s the former Land Survey Board and the Geographical Survey Office acquired their first flatbed plotters and started production and development work related to topographic mapping. Now a number of governmental authorities as well as private enterprises make quite extensive use of computer-controlled methods in their cartographic production.

Much work has been devoted to the development of software and methods for computer-aided topographic cartography in Sweden. Thus the system now used by the Land Survey was almost completely developed by this agency (Ottosson, 1974). The system is based on the use of a Prime 400 computer system (512K core, 400 megabyte disk) with FORTRAN as the programming language. The three biggest cities, Stockholm, Gothenburg and Malmö, are jointly developing a software system which is used for integrated data processing of local surveys for map production on flatbed plotters. Development work is also carried out by some private enterprises and by the Royal Institute of Technology in Stockholm. The work carried out at this institute aims at computerizing photogrammetric large-scale mapping as well as at establishing a basis for the education on computer-aided topographic cartography which is mainly located in this institute.

LARGE-SCALE TOPOGRAPHIC MAPPING

Large-scale mapping is, as mentioned, a responsibility of the communities. However, the main part of the production of maps is carried out by private enterprises and by the Land Survey as a commercial service. Some of the large cities have introduced digital methods in map production.

The following section reviews the situation as of 1979 regarding computer-aided large scale topographic mapping.

The Land Survey

To some extent the problems of computer-assisted production are fewer for large-scale mapping than at smaller scales. Thus problems like generalizing do not exist in large-scale map production. On the other hand it is, of course, to a large extent possible to apply the same technique in both large and small-scale mapping. Thus the development at the Land Survey of software and methods for computer-controlled large-scale mapping is very much linked to development for small-scale topographic map production. Part of the applications typical for large-scale mapping are presented here.

A special software system called IDAK has been developed to be used in integrated

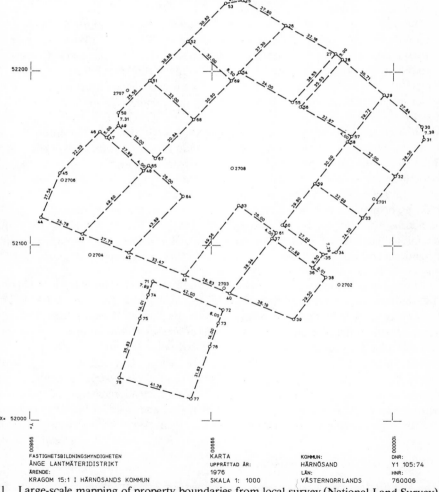

6.1 Large-scale mapping of property boundaries from local survey (National Land Survey)

data processing of detailed survey measurements and numerically controlled plotting of measured details. This software is primarily used in routine survey work at the local offices of the Land Survey, especially in work concerning real estate formation. An example of the use of the IDAK program is given in Figure 6.1. As can be seen from this example, the software used for this purpose also performs a kind of computerized editing in such a way that errors such as the duplication of point numbers are avoided.

In all kinds of photogrammetric plotting, accurate base sheets are needed in the compilation. The construction of base maps is a procedure well suited to the computer. Thus, one of the first programs developed for generating control information for computer-aided drafting was a special application program used in base sheet plotting.

This software is mainly used in the production of base maps for large-scale photogrammetric maps. The customers have, for many reasons, varying requirements regarding the layout of the base sheets. This fact has made it necessary to build up the program in such a way that the many different demands can be met. Thus, when using the program a number of input parameters must be specified. In order to facilitate use, the parameters defining the layout of standard base maps are fixed in the program.

Photogrammetric methods are used to a very large extent in the production of large-scale maps. In order to rationalize this map production, much work has been devoted to the development of software and techniques for a semi-automated photogrammetric map procedure. In this technique planimetric details and (as an exception) contours are digitized in the stereo instrument. The software used for the plotting of map details allows map drafting in various layouts to meet the different standards used in large-scale mapping in Sweden. The initial work with this technique has shown great advantages, especially regarding mapping of highly urbanized areas at map scales of 1/400 or 1/500. An example of such a map is shown in Figure 6.2.

The software used in this semi-automated photogrammetric mapping is based on a data base management system called KASOF developed at the Land Survey. This system includes software for reading and writing data in the data base. Different kinds of data in the base are stored in various addresses according to type. Examples of such address types are points, open or closed polygons, and text-strings.

The different addresses of the data base are directly accessed by three different keys: location, category, and status. Furthermore, all addresses can be read sequentially. Using, for instance, the location key all addresses of the data base within a specified area are accessed. Analogously all addresses with a given category or/and status code are easily retrieved from the data base by a direct access method.

Private Enterprises and Communities

The VIAK (Via et aqua) consulting firm is the most advanced enterprise in Sweden

6.2 Large-scale map scribed on a Kongsberg flatbed plotter (original scale 1/400). (National Land Survey)

as regards development in C.A.C. There, a software system called 'DIGIKART' is under development. This software makes use of a specially designed data base management system to take care of large-scale mapping problems. The system includes a Nova 3/D minicomputer, a Kongsberg 1215 flatbed plotter and a number of terminals for input and output of data.

Furthermore, VIAK has developed and introduced in practical work a mini interactive graphic system for computing and plotting based on the use of a Tektronix desk computer. A Kongsberg 1213 flatbed plotter is connected on-line to the desk computer. This system, which is comparatively inexpensive, has been designed to take care of part of the mapping activities in local offices.

Some of the larger communities have implemented flatbed plotting systems, mainly of the Kongsberg type, in their large-scale map production. This is the case for instance in Gothenburg, Malmö, Oerebro and Botkyrka. Gothenburg was the first Swedish city to introduce computer-controlled map drafting. Recently, a Computer Vision interactive graphic system has been added to the equipment for computer-assisted map production in Gothenburg. The situation in Malmö is almost equivalent to that in Gothenburg.

The plotting systems are used for different large-scale mapping purposes. One main task in some of the communities is revision of 1/400 maps. In this task some planimetric details are measured on existing old maps. These details and coordinates of new planimetric details determined by local surveys are stored in data bases. Flatbed plotters are then used to produce originals for new 1/400 (or 1/500) primary maps. Still another application is production of overlays showing sewage systems as well as different kinds of cables and pipelines.

SMALL-SCALE TOPOGRAPHIC MAPPING

As already stated, the responsibility for the official topographical mapping rests with the Land Survey. Thus the main activities in Sweden concerning the computerization of map production at scales of 1/10,000 and less take place in this agency.

The main reason for introducing computer-aided techniques in small-scale topographic mapping at the Land Survey has been to eliminate as far as possible manual procedures in order to reduce costs. However, other reasons do exist. One of these is the fact that map consumers need landscape information not only in the shape of maps but also in digital form in order to take advantage of EDP in different planning and computing activities.

Already initial studies of the possibilities of introducing C.A.C. have shown that replacement of all manual procedures in the production of small-scale topographic maps by computer-controlled techniques is not economically justified. A main reason for this is the fact that digitization work, at least for the time being, is a very expensive procedure especially if one takes into consideration all work and costs related to correcting and completing data.

As a result of the problems and costs related to the implementation of computer-aided techniques in topographic map production, it has been found that step-by-step changes from manual to computer methods are preferable. This philosophy is still relevant and it is quite clear that the character and dimension of the technical and economic problems involved in C.A.C. are such that a step-by-step approach is fully justified.

The following section gives a review of the present state of C.A.C. applications at the Land Survey.

Computer-Aided Scribing

All official maps are supplied with at least one reference grid and as a matter of fact the Economic Map as well as the Topographic Map contain three different reference systems.

For many reasons the subject of the first attempts to computerize cartographic drafting work was just the construction of reference grids. Thus in 1968 the first computer program was written to generate control tapes for a Konsberg plotter. The program was rather simple and did not permit automatic plotting of all kinds of complex grid constructions used on various official maps. However, the work carried out by this program did show that it was reasonable to use automatic plotting for this part of cartographic drafting. Therefore, a new general computer program capable of generating control tapes for automatic drafting of reference grids, frame information etc. with a very flexible layout was developed (Ottosson and Tönnby, 1973).

The present version of the program contains slightly more than 8000 FORTRAN statements and is divided into some 100 subroutines with special care put on the structure to permit the addition of new facilities.

The most common use of this so-called GRID program is production of grids for maps at the scales of 1/10,000 to 1/100,000. The computer-controlled scribing of grids includes scribing of coordinate numbers, UTM coordinates and frame lines.

The economic saving of using flatbed plotters in scribing reference grids is significant. Thus, in scribing the three different grid originals used in the topographic map at 1/50,000, the cost reduction has, for instance, been estimated at about 75 per cent. Maps at scales of 1/250,000 or smaller often contain grids with a more complex layout. In such cases automated scribing of grids becomes even more economic.

Another example of the application of flatbed plotters in the construction of reference grids is the plotting of so-called Decca-grids. The Decca system is an internationally used navigation system at sea as well as in air traffic. The mathematical equations defining the so-called Decca lanes in a certain map projection are relatively complex. Thus manual construction of Decca grids is a very time-consuming job. Therefore a special application program for computing control information for automatic plotting of Decca grids has been developed.

The formula defining the Decca lanes in a map projection system are so complex that it is necessary to compute lane values for intersection points in a regular grid and then use interpolation techniques to compute the separate Decca lanes. The program developed for the computing of Decca lanes also takes care of the numbering of the lanes. An example of a Decca grid is shown in Figure 6.3.

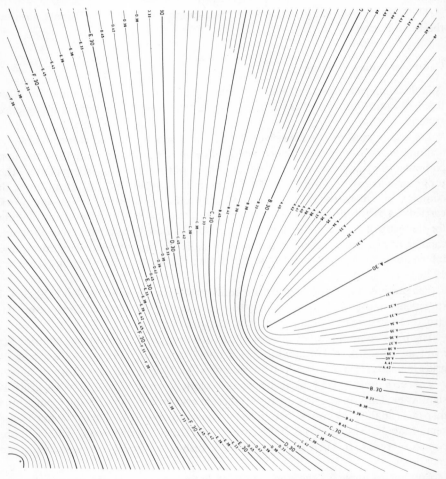

6.3. Decca lanes plotted and numbered by scribing on a Kongsberg plotter (National Land Survey)

Real estate boundary lines, as well as boundaries of villages, parishes, communities, countries, etc. are commonly shown on maps by using different symbols. The symbols are normally built up of dashes, dots and crosses of varying sizes.

From a cartographic point of view two main conditions must be fulfilled in computerized drafting of border lines. First, it is important that points defining border lines are clearly marked on the map. For this reason it is, for instance, not possible to use any hardware dash-line function of the plotter in drafting of dashed border lines. Thus software has been developed to take care of this problem for the great number of different combinations of symbols that are commonly used to denote border lines on all kinds of topographic maps. Secondly, when digitizing is performed to determine coordinates for border line points it is important that points measured more than once are given an unambiguous position. Unless this is taken into consideration, lines passing through the same point on the manuscript do not intersect. This can be corrected by interactive techniques but a better solution is to store all border line point coordinates in a buffer. Then if a new point is situated within a given distance from a point stored in the buffer, the new point is replaced by the buffer point.

Computer-aided techniques are frequently used in scribing boundaries on originals of the economic map as well as the topographic map. The technique is particularly used in areas where the same information can be used not only to produce the land-use map but also to produce at the same time a map at the scale of 1/5000 which is needed for certain planning and registration purposes.

Aeronautical charts are subject to relatively short updating intervals. The aeronautical information is normally published as an overprint on unaltered base maps. To a relatively large extent it has been easy to computerize the production of these overprints. For this reason aeronautical information has been the object of numerically controlled drafting for many years. To solve this problem an aeronautical information system has been developed by the Land Survey in cooperation with the Swedish Air Force. The aim of this development has been to create a system which at all times contains current aeronautical information concerning Sweden as well as parts of the neighbour countries. An application program for computing of control information to Kongsberg plotters is part of this data bank system. Typical plotting scales are 1/250,000 to 1/6,000,000.

Light Exposure Techniques

For several reasons scribing techniques are commonly used in production of small-scale map originals. It has been possible to apply the same technique in computer controller plotting since the Kongsberg plotters used for this purpose are equipped with tangentially controlled scribing tools. However, to some extent it has been advantageous to make use of a light exposure device. This is the case when map symbols have to be plotted.

Thus a light exposure device is used at the Land Survey in small-scale topographic mapping to expose all map symbols on light-sensitive material. Coordinates for the symbols in question are measured on the map manuscripts which are normally

1 T ÖVERSIKTSKARTAN

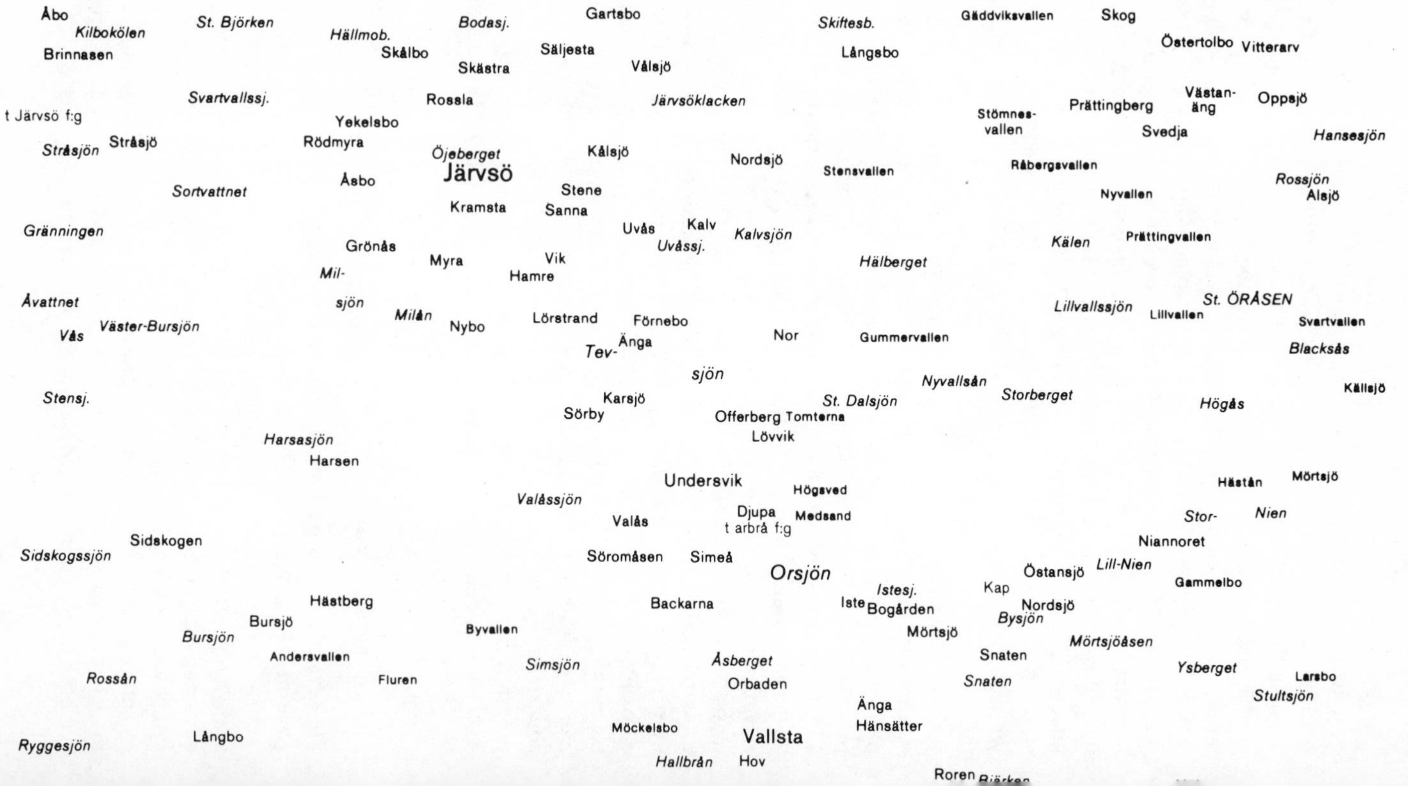

6.4 Place names for a 1/250,000 map (text produced by computer-controlled photo-typesetting) (National Land Survey)

produced at a larger scale than the final map. Some of the map symbols, such as houses, are given an appropriate rotation on the map. This makes it necessary to measure not only one point to define the exact location of the symbol but also an additional point in order to define the rotation of the map symbol.

The experience from the use of photo exposure of map symbols is very good. The digitizing of map symbols is a relatively quick procedure and the light projector also works very fast. As an average for a sheet of the topographic map at the scale of 1/50,000, 8 h is needed in the flatbed plotter for the light projector work. This figure is to be compared with some 75 h which are needed for manual scribing of the same symbols. This means an 80 per cent reduction of man-hours compared to manual scribing of the symbols. The quality of the originals prepared by light exposure is very high and exceeds by far the quality reachable by manual scribing.

Computer-Aided Photo-Typesetting

All topographic maps contain text to some extent. As a matter of fact in an organization such as the Land Survey a great deal of time is spent on providing map originals with adequate text. Hitherto all this work has been carried out with pure manual methods. In order to rationalize this part of the map production, a computer-controlled photo-typesetter of the Diatronics type has been acquired. This instrument has quite a high capacity and is able to set about 30,000 characters per hour. Eight different type founts can be stored at the same time in the instrument and are easily interchanged with simple orders. Typesetting is performed on photographic film with a maximum size of 30×30 cm.

From a cartographic point of view it is of great importance that such an instrument allows the possibility of typesetting in predetermined positions. Thus, if coordinates are determined for geographic names and other text material, the text pressing can be carried out automatically by the photo-typesetter. To make this possible, a software system for generating control information for the photo-typesetter has been developed at the Land Survey.

The first steps towards a more complete automation of the text pressing has been taken. Experiments and practical work have shown that computer-aided text-setting means quite a substantial rationalization of this work. This is particularly true in complete map revision when the whole text content has to be changed. Figure 6.4 gives an example of computer-controlled typesetting used in revision of the text original of the general map at the scale of 1/250,000.

The use of computer-controlled photo-typesetting in topographic mapping has several advantages. Apart from a relatively great reduction of manpower, the fact that all names are stored in a data base is of great significance. This data base will certainly be of great importance in future map revision, in derived mapping as well as in production of gazetteers and so on.

Digital Terrain Elevation Data Bases and Computer-Controlled Orthophoto Mapping

Low-Density Terrain Elevation Data Base

Information concerning elevation conditions of the landscape is certainly of great interest in many activities. Hitherto, information regarding this topographic feature has normally been presented on maps as contours in combination with spot elevations for certain points of interest.

In Sweden the first demands for digital terrain elevation data arose some 15 years ago. The prime reason for this was the need for digital data in connection with computer-aided methods for radar visibility studies performed by defence authorities. Thus data capture in order to establish a digital terrain elevation data base was started at the Land Survey in the middle of the 1960s. At that time, purely manual methods were used in the data acquisition, which was carried out using the topographic map at the scale of 1/50,000 as base material. In this map series spot elevations were manually interpolated for points in a regular grid with an interval of 500 m. Such elevation data is now available for about 80 per cent of the country.

Terrain elevation data with such a low density certainly has a restricted value not only in radar visibility studies but also in many other applications. Therefore a demand for denser information has arisen during the last few years. For financial reasons it has not until recently been possible to fulfil these wishes for higher density. Now the introduction of numerically controlled orthophoto production has created new conditions for the realization of a digital terrain elevation data base containing information with higher density.

Computer Controlled Orthophoto Production

At the time when the production of the economic map started in the middle 1930s the photo map was produced as so-called controlled mosaics. This method was used until 1966 when the first Gigas–Zeiss orthophoto projector was introduced in the production. Since then the production of the economic map has been based on orthophoto maps.

Some years ago computer-controlled orthophoto projectors appeared on the market. Realizing the great advantages offered by such systems, the Land Survey started experiments and software development in 1976 in order to study the possibilities of introducing numerically controlled orthphoto production. These experiments indicated great advantages, particularly regarding image quality, flexibility, and economy, and resulted in the acquisition of an Avioplan OR1 orthophoto projector system manufactured by the Wild Co., Switzerland. This equipment was introduced into regular production in 1978.

When using the Avioplan equipment it is necessary to have access to digital terrain elevation data with a high density. It is, of course, possible to acquire control information for each photo to be projected but in order to take full advantage of the great flexibility inherent in the computer-controlled projection system it seems to be very rational to establish a high-density terrain elevation data base to store adequate information in a suitable form.

Since orthophoto production is of very great importance in Swedish cartographic activities and is expected to be of even greater significance in the future, especially for map revision, the implementation of computer-aided methods in orthophoto production has led to the establishment of a high-density terrain elevation data base (Ottoson, 1978b). This decision has, of course, also been influenced by the fact that the data base certainly is of great interest for many applications other than cartographic.

High-Density Terrain Elevation Data Base

Three methods are used in capturing basic information for the data base:

    photogrammetric measurements of aerial photos;
    digitizing of existing contour plates;
    digitizing of existing profile plates.

Photogrammetric data acquisition can be performed in the profiling or contouring mode. In orthophoto production profile measurements are often carried out to capture elevation data from aerial photos. One great advantage with this method is the fact that the time needed for the data capture is relatively independent of the terrain elevations, which is not the case if data are acquired in the contouring mode. For this reason, profile measurements are used to a very large extent in data acquisition from aerial photos.

Over much of Sweden, high-quality elevation contour plates have already been produced and this material can be used as a base for data acquisition.

For the time being there are no automatic devices available in Sweden which can be used to digitize existing contour plates. Automatic raster scanners in which a contour plate can be digitized very rapidly do exist, but this digitizing technique calls for very extensive and expensive data processing. Furthermore, since the computer cannot be programmed to identify an individual contour line, an interactive graphic system must be used for this purpose. Thus at present, the use of ordinary digitizing tables for data capture from elevation contour lines seems to be the most realistic solution. As Boyle has indicated in Chapter 4, recent technical advances may change this.

Since this kind of data acquisition is relatively expensive, the method can only be used in comparatively flat areas. Thus in areas with more hilly terrain it is at present more economic to use other techniques for gathering the data.

As regards the third main method for data capture, digitizing of existing profile plates, it has already been mentioned that orthophotos have been produced for large parts of the country. Since Gigas–Zeiss projectors have been used in this production in the so-called off-line mode, glass profile plates have been used to store terrain elevation data in an analog form. Thus a great number (about 14,000) of these plates is available.

It has been proven possible to use the profile plates in elevation data acquisition and for this purpose one of the Gigas–Zeiss projectors has been equipped with shaft encoders and devices for output of digital coordinate data. This method works very well, and of special interest is the fact that a profile plate can be recorded in less that $\frac{1}{2}$ h. Our experience is that digitizing of existing profile plates is, in most cases, the fastest and most economic method for acquisition of digital elevation data. Thus, the fact that existing profile plates cover about 60 per cent of the country is of great economic importance in the establishment of the new terrain elevation data base.

The choice between the three main data acquisition methods is dependent on existing materials (contour plates and profile plates) and their quality as well as the conditions of the terrain and other natural features. A total of about 400,000 km² are planned to be covered by this type of data base which corresponds to about 90 per cent of the country. Only the mountainous areas of the northern part of the country will not be included.

The most frequently used technique will be digitizing of existing profile plates. This method will be used for about 220,000 km². Digitizing of existing elevation contour plates will be carried out for some 100,000 km² while the remaining 80,000 km² will be captured by photogrammetric measurements of aerial photos.

A total of about 50,000 h will be needed to capture digital terrain information for the entire data base. The creation of the new data base started in 1978. The data acquisition will be carried out in accordance with a plan which takes into account the needs for digital information for production of orthophoto maps as well as for other activities. Due to the close connection between the production of orthophotos and the data base, it is estimated that data acquisition will not be completed until country-wide coverage of orthophoto maps has been attained. This will take approximately another 10 years.

Once basic terrain elevation data has been acquired the information has to be fed into a computer system for processing and storage.

Many alternatives exist which can be used for storing information in the data base. The Land Survey has, for several reasons, chosen to store elevation data in the form of spot elevations in regular grids. The sheet division adopted in the production of the economic map at scales of 1/10,000 or 1/20,000 is used as a base for the high-density elevation data base with a grid size of 5×5 km or 10×10 km respectively. Thus the grid interval is 50 or 100 m respectively.

Since data capture in only a very few cases is performed for grid intersection points, it is necessary to transform measured values into grid points with interpolation techniques (Akima, 1972; 1974).

Each grid of the data base contains 101×101 or 10,201 points. The final number of grids which will be stored has been estimated at 13,000, which corresponds to a total of about 130 million points.

As already stated, the prime reason for the establishment of the terrain elevation data base is the production of orthophoto maps. However, digital elevation data is of great interest in many other applications. The following applications can be mentioned:

- computing of cross-sections, i.e. for visibility studies;
- correction of remote sensing data;
- correction of gravimetric measurements;
- terrain correction in flight simulator systems;
- different cartographic applications such as slope mapping and automated hill shading;
- military applications such as computer-controlled navigation in aircraft and guided missiles.

The elevation data base will certainly rationalize the production of official maps especially by making the orthophoto process more flexible. Apart from this, the examples given above indicate that the data base will be of great value in a variety of planning and computing activities.

PART TWO: COMPUTER-ASSISTED THEMATIC MAPPING

BACKGROUND

The first step in computer-assisted thematic mapping in Sweden was taken in 1955 by Professor Torsten Hägerstrand at the University of Lund. In an article in the *Swedish Geographical Yearbook* (Hägerstrand, 1955) he outlined the methodology by which every person in the population file of the taxation authority could be referred to the building where they lived and the results portrayed on the economic map of Sweden at the scale of 1/10,000. This map covers most of Sweden and has a coordinate system. By recording the coordinates for the buildings and data from the taxation and census file in a study area he was able to produce the basis for population maps. This was one of the earliest examples of C.A.C. in the world.

Since the records in the population file do not contain the identification of the house each person lives in, but only the designation of the real property, Hägerstrand studied the loss of precision and only a central point was used for location purposes. He found that 'the coordinates for such an identifying point should be a same mandatory appendix to the property designations as the civic number to the person's name' (Hägerstrand, 1955). He also found that a list of the coordinates of the real properties was a great help when locating properties on maps.

Hägerstrand's study led to the initiation of several research projects supported mainly with funds from the Swedish Council for Building Research. Government responded to Hägerstrand's findings by proposing the registration of coordinates for all real properties in Sweden. Research and development in this area has expanded in several directions since Hägerstrand's initial innovative proposals.

DATA SOURCES FOR THEMATIC MAPPING

Population Registers

The population registers are kept in different forms and at different levels. The parish ministers act as public officials in regard to registration and keep books with various data, such as births, deaths, marital status, and changes of residence. These books have existed for about 250 years. The actual data for each individual are kept on a card file which still constitutes the official population register.

When the parish minister changes information on a card he also sends a copy of it to the data centre of the county adminstration where records are computerized by using personal identification numbers as a key. The provincial registers are updated once a week using computer technology. These registers are used mostly as a basis for further special registration such as taxation and election lists, in information systems at the county level, and for further updating of registers at the national level. Such registers are kept for social welfare, military service liability, statistical, and other purposes.

For statistical purposes, the Central Bureau of Statistics also keeps some hundred computerized population registers with sub-groups of the total population. Even hospital and private firms such as banks and insurance companies keep population registers. In all these registers each person is listed by personal identification number. This means that both updating and exchange of information between registers are very easy.

In the provincial population register at the county administration and in the national register at the Central Bureau of Statistics the location of each person's residence is described both by the identification number of the real property and the postal address of the residence.

Register of Business Firms and Farms

Data on business firms are also kept in registers at different levels. The basic data capture is carried out and maintained by the local taxation authority in each municipality. These data are registered in machine-readable form at the county administration for taxation purposes. All firms obliged to render an account of value added tax are recorded in an additional special register.

With these registers as a basis, together with information from the population registers for taxation and health insurance, the Central Bureau of Statistics has built a central register of business firms. This register is mainly used in inquiries on topics such as unemployment.

At the end of 1978 a governmental investigation proposed that the central register of firms in connection with the 1980 census should be extended to a register of all work sites. The government has presented this proposition to Parliament. The location of each work site is to be determined by the postal address. The Central Bureau of Statistics has also been asked to further investigate the possibilities of linking information in the register of work sites with information in the population register. If this is successful, information on an individual level for both place of work and place of residence can be obtained. This gives new and extremely interesting opportunities for the analysis of flow data and other types of spatial interaction.

A register of farms is also kept at the Central Bureau of Statistics. The location of each farm is given by the identification number of the real property which constitutes the farm.

Both the register of business firms and the register of farms are confidential. The use of the registers is restricted to administrative purposes, but with special permission they can be used for research and other purposes if it can be guaranteed that no business information about individual units can be derived from the statistics.

Property and Land Register

As the previous section has indicated, the real property plays an important role as a basic areal unit. Even in the legislation for building, environmental protection, nature reservations, and other obligations and restrictions on the use of land the area of the actual regulation is described in terms of a list of one or more real properties.

Historical reasons explain why the information on real properties has been divided into two registers. The property register is the oldest one and contains information such as how the property was created and its area. The land register, on the other hand, contains the title of the land, mortgages, easements of an economic nature, leaseholds, and other related information.

In 1968, Parliament decided to carry through a reformation of the registration of real property data. The implementation of this decision has been controversial, but the basic idea of converting the manual registers into machine-readable form is still valid. The Central Board for Real Estate Data (C.F.D.) was given the responsibility for the implementation of this process.

The Register of Coordinates

The 1968 Act of Parliament also contained a decision to record coordinates for each parcel of all real properties in Sweden. For C.A.C. and spatial analysis this created an incredible opportunity to derive locational determined data.

The basic requirements for a geocoding exercise of this type are the existence both of adequate technical equipment and base maps of good quality and suitable scale. Good digitizers were not available at that time so C.F.D. had to order more of the specially designed prototypes which had been used in previous investigations.

The maps which serve as the register maps in the property register are of different qualities. In rural areas the register is based on the economic map at the scale of 1/10,000. This map is produced on an orthogonal coordinate system. Each map sheet covers an area of 5 × 5 km. In regions with many small properties, sheets covering areas of 1 × 1 km are produced at the larger scale of 1/2000. Since it has taken more than 30 years to finish the economic map for the whole of Sweden great differences in quality exist between the oldest and the newest sheets.

In urban areas, local authorities are usually responsible for the base maps. This means that some maps are extremely good, while others are more illustrations than true maps. Using poor maps when recording coordinates often means that redigitization has to take place when new and better maps are produced. This substantiates the importance of good base maps emphasized by Harris in Chapters 5.

After some experimental work and technical development of both hardware and software, C.F.D. began, in 1969, to create a coordinate register as a sub-register to the forthcoming computer-based property register. C.F.D. first requests the property registration authorities to provide copies of the property base maps.

At C.F.D. the maps are further prepared before geocoding takes place. Some area codes are inserted and the map sheets are put edge to edge to record on which sheet the central point of properties covering more than one sheet has to be located.

The digitizing is now performed in on-line mode with interaction between the operator and a minicomputer. The operator writes the alphanumerical property designation and the type of the registered point as follows:

C   centroid;
B   building point;
CB   centroid when it is known that there is a building close to it;
BC   building point in the centre of the property, no central point registered;
R   ancient monument with identification;
K   edge point;
BG   group of buildings.

Special attention during registration is paid to routines designed to give as correct data as possible during the initial registration process. The need for accuracy in the location of the coordinates is not very high. The hardware used has a resolution of about 1 mm. In general, at the input map scales used, that means 10 m in rural areas and 2 m in urban areas. Since many register maps are of very low accuracy themselves it is difficult to obtain true locational accuracy when digitizing points from these maps.

After digitizing, the data are processed in a main frame to create a preliminary coordinate register. This register is manually checked against the register books, which contain several real properties not shown on maps. In many cases, extensive research is required to find out if the property really exists or not, and to locate it on a map in order to give it coordinates.

During the last few years the resources available for the process of registration of coordinates have been reduced while at the same time the task of updating has become heavier. There are about 217 million real properties in Sweden. By the end of 1978 half of them had been registered. It is planned to complete registration for the whole country by 1986.

The property registration authorities are responsible for the daily maintenance of the register. The registrar fills in updating forms and sends them monthly to C.F.D. The most crucial thing is to keep the building points up to date.

METHODS FOR HANDLING COORDINATE-BASED DATA

After Hägerstrand's introduction of the so-called 'coordinate method' in 1955, it was evident that further research and development was needed. Among the first problems which had to be solved was how to identify both points and polygons and how to find a representative point for the location of polygons.

Nordbeck (1962) found that a central point for a polygon could be defined as the centre of the largest inscribed circle or the centre of the smallest circumscribed circle. Using the latter, however, has the disadvantage that in some circumstances points can be located *outside* of the polygons. Since these are used to calculate the centre of the largest inscribed circle it is not possible to use this approach when digitizing central points of polygons. Some approximation therefore has to be used. Nordbeck found that for real property areas the base point of the property identification number was a very good approximation of a mathematically defined central point. Later on, in fact, this point was used by C.F.D. The point-in-polygon problem was also considered and given a general solution by Nordbeck and Rystedt (1967).

Basic Problems in Mapping

Hägerstrand's first maps were square grid net maps. These maps are also called chorological matrices and can be obtained by cross-tabulation in respect to the x and y coordinates. 'The square grid net map is the simplest map which can be produced' (Nordbeck and Rystedt, 1972b).

Isarithmic Maps

A more widely used map, especially among geographers, is the isarithmic map. An isarithm is a line which connects points of equal value. Constructing isarithmic maps is equivalent to graphically displaying the function $z = f(x,y)$. In this instance, for example, z could be the height above sea level at the point with the coordinates x and y. It has been argued that it is not meaningful to construct isarithmic maps if the function is a discontinuous one, and that it is incorrect to make such maps showing population density if they are based on point-referenced data for polygons. These objections are valid if the central points of the polygons are used as data points in a simple interpolation. By using the overlapping technique, however, this obstacle can be overcome.

The manual method to obtain overlapping is the principle of the floating circle. Let the circle be placed with its centre at the arbitrary point (x,y) and z be the total number of objects within the circle obtained by summing the data values for all data points in the circle. In this way, a function $z = f(x,y)$ is defined at every point. It has been shown (Nordbeck and Rystedt, 1970) that functions based on reference areas such as the floating circle and overlapping technique have continuous qualities. Hence, it is theoretically proved that it is correct to construct isarithmic maps based on point-referenced data. An American study (Kolberg, 1970) arrives at the same conclusion.

The usual method used when constructing isarithmic maps is to calculate the function value in the grid points in a regular grid. If the distance between the grid points is denoted by H and diameter of the reference area by D the overlapping constant S is defined as $S = D/H$. Overlapping is obtained when S is greater than 1. A suitable value of S is 2 (see Figure 6.5). Furthermore, the reference area has to be greater than the largest polygon when data based on point-referenced polygons are used. Many bad maps have been produced when these two simple rules have not been followed.

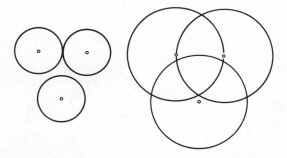

6.5　Three grid points in a regular grid. (In the smaller diagram no overlap is shown ($S = 1$) In the larger, overlapping results in a value of $S = 2$)

A commonly used method to obtain the function value of a grid point is to use the values of the closest data points weighted in accordance with the distance to the grid point. This method should only be used when the values of the data points can be considered as observations of a continuous function.

Dot Maps

The main purpose of a dot map is to give a visual display of the geographical distribution of various phenomena. For this purpose such maps can be very useful. The greatest drawback at present is the difficulty of producing dots on the plotter which are suitable for reproduction. Another problem is where to locate the dot when it represents more than one unit. C.F.D. has chosen the gravity point of the units the dot represents (Olsson and Selander, 1971).

Shaded Maps

Numerous programs for producing shaded maps have been made all over the world. In Sweden, a method to produce shading in colour was developed in the mid–1960s. An apparatus for copying pictures in colour was built at the Lund Institute of Technology. The picture to be copied was put on a rotating drum. A sensor on a lead screw parallel to the drum scanned the picture and sent signals to three ink jets with the colours red, blue and yellow. The ink jets were placed on a lead screw similar to that of the sensor, and directed to a sheet of paper on another rotating drum. By giving the ink drops in the jet streams electrical loadings they could be manipulated and mixed into the same colour as the original picture.

When the origin of the signals to the ink jets was changed into three binary matrices, one for each basic colour, in a computer memory, the colour plotter was invented (Bergström, Jern, and Wissler, 1977).

Along with the development of the plotter a versatile software system was developed (Jern, 1975). In Figure 6.6 a reproduction of a colour plot is shown in black and white.

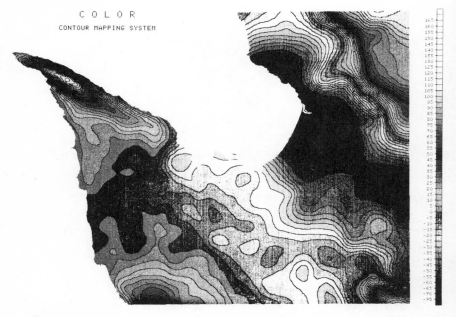

6.6 A Reproduction in black and white of a contour map produced by the colour plotter (Lund University)

RESEARCH AND APPLICATIONS IN SWEDEN

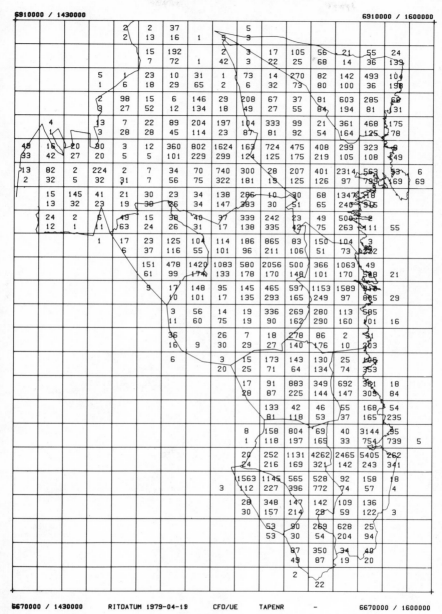

6.7 A square grid net map produced at C.F.D. (The upper figure in each square shows the number of permanent homes and the lower the number of summer cottages.) The grid size is 10 × 10 km. The coastline and municipality boundaries are selected from a computer-stored digital map of Sweden

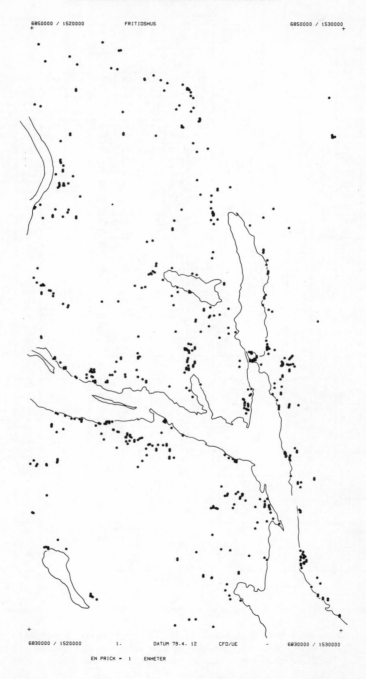

6.8 A dot map showing an enlargement of an area of Figure 6.7 (Each dot represents one summer cottage. Where there is more than one cottage on one real property the dots are regularly spread around the centroid of that property)

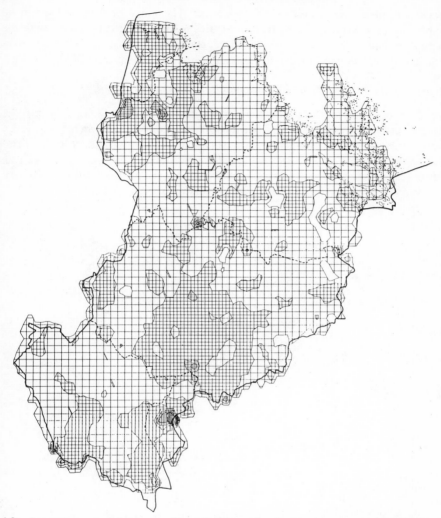

6.9 Average personal income in Uppsala County (an isarithmic map produced at C.F.D. Income increases with shading density and the distance between grid points is 2 km. The reference area is a circle with a diameter of 5 km. Thus the overlapping is equal to 2.5; cf. Figure 6.5)

EXAMPLES OF ACTUAL MAPPING
Mapping in the Land Data Bank System

During the last 7 years considerable effort at C.F.D. has been devoted to developing data and computer programs suitable for C.A.C. on a commercial production basis. Now C.F.D. can offer maps of high cartographic quality to physical, economic, social, and other urban and regional planners. The range of data which can be mapped is quite large as can be seen in Figures 6.7 to 6.9. In cooperation with the Central Bureau of Statistics, it is planned to enlarge this service on the basis of data which will be collected in the population and housing censuses of 1980.

Mapping at the Swedish Meteorological and Hydrological Institute (S.M.H.I.)

The daily need for maps for weather forecasting has made S.M.H.I. possibly the biggest producer of maps in terms of numbers in Sweden. S.M.H.I.'s computerization of map production started in 1968 with use of ordinary line plotters. The step to produce only computer-plotted weather maps was taken in 1977. This was possible because the electrostatic plotter reduced the plotting time considerably (Lönnqvist, 1978). In Figure 6.10 a weather map is shown.

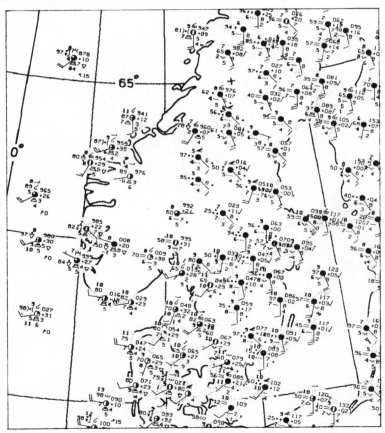

6.10A Weather observation in symbol form at 10:00 a.m., November 1, 1977 (Lönnqvist, 1978, p. 91). Reproduced by permission of Liber Kartor, Stockholm

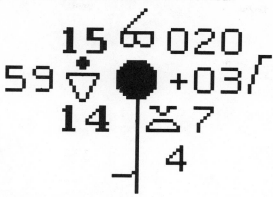

6.10B Enlargement of a single weather observation plotted by an electrostatic plotter (Lönnqvist, 1978, p. 92). Reproduced by permission of Liber Kartor, Stockholm

The observations from the weather stations are not only plotted; they are also calculated at grid points in a regular grid net and with rather complicated methods used for computer production of forecast maps. These maps are usually displayed as isarithmic maps showing air pressure and temperature. The isarithms are located by using the values for sixteen grid points and interpolation in a cubic spline funtion. This gives soft curves and reliable gradients, which is important when dealing with meteorological problems.

Mapping in Regional Development Research

In the 1960s a governmental group of experts for research in regional development (E.R.U.) was set up. Computer mapping has been an essential part of the studies produced by the group during the last decade and has played an important role in the analysis of regional development (Hertling, 1977).

As the researchers at E.R.U. work at the national level they can use more aggregated data. By special processing of the population and census registers, appropriate variables are created and aggregated for the parish level. The regional code for parishes is then used as the identification when these variables are matched with a coordinate register with the central points for the parishes. The geographical distribution of the variables has usually been displayed as isarithmic maps. In Figure 6.11 a series of such maps is shown.

OTHER RESEARCH

There is, of course, much other research in computer-assisted thematic cartography in Sweden. The Department of Geography at the University of Stockholm makes extensive use of the COLOR plotter in the interpretation of satellite data. The Bolinden Mining Company is also using this plotter to display and analyse prospecting data.

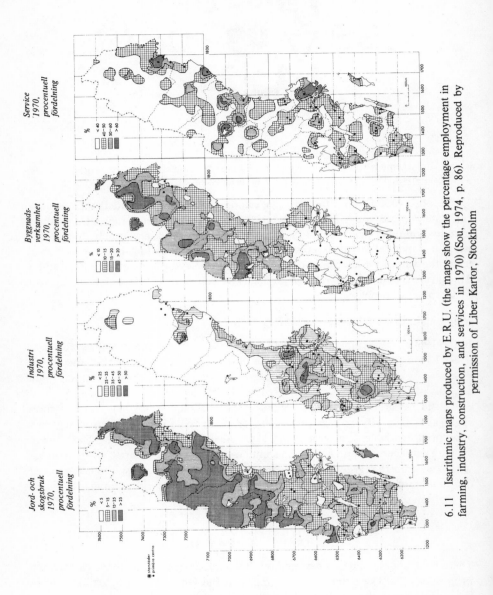

6.11 Isarithmic maps produced by E.R.U. (the maps show the percentage employment in farming, industry, construction, and services in 1970) (Sou, 1974, p. 86). Reproduced by permission of Liber Kartor, Stockholm

CONCLUDING REMARKS

Computer-assisted mapping of statistical data in Sweden is at a critical point. Swedish researchers were pioneers in this field and many years of development and research have resulted in the production of useful and economic maps. However, if these maps are not effectively used then support for research and development is likely to decline or even be withdrawn. There is a strong movement in Sweden which sees computer technology as a threat to privacy and individual freedom, and computer-assisted thematic mapping is dependent upon access to information in machine-readable form.

The utility of maps to the planner and to the public is not as readily obvious as cartographers like to think. C.A.C. has speeded up map production and made the map a much more powerful analytical tool, but acceptance and use of this innovation has been slow and much more information-diffusion and education is required to ensure that the considerable potential is realized. In this respect the future may well lie more with the interactive map on a C.R.T. than with the traditional hard copy.

In computer-assisted topographic mapping considerable progress has been made but a sensible and balanced approach to the mix between computer and manual methods still holds out most promise for the future.

REFERENCES

Akima, H. (1972). 'Interpolation and smooth curve fitting based on local procedures', *Communications to the ACM*, **15**, 914–18.

Akima, H. (1974). 'Bivariate interpolation and smooth surface fitting based on local procedures', *Communications to the ACM*, **17**, 18–20.

Bergström, L. A., Jern, M., and Wissler, B. (1977). 'Experiment med data i färg', Swedish Council for Building Research (S 12: 1977), Stockholm.

Hägerstrand, T. (1955). 'Statistiska primäruppgifter, flygkartering och "Data Processing"—maskiner. Ett kombinationsprojekt', *Svensk Geografisk Arsbok 1955*, 233–55.

Hertling, H. (1977). 'The ERU mapping system', *Cartographica*, Monograph No. 20, Supplement to *The Canadian Carographer*, **14**, 59–67.

Jern, M. (1975). *COLOR: Software for Hard Copy Color Display System*, Lund University Computer Centre.

Kolberg, D. W. (1970). 'Population aggregations as a continuous surface: an example of computer mapping', *The Cartographic Journal*, **7**, 95–100.

Lönnqvist, O. (1978). Kartframställning vid Sveriges meterologiska och hydrologiska institut. *Sverige kartläggning, tillägg 1968–1977*, pp. 86–92. Kartografiska sällskapet, Liber Kartor, Stockholm.

Nordbeck, S. (1962). 'Location of areal data for computer processing', *Lund Studies in Geography*, C2.

Nordbeck, S. and Rystedt, B. (1967) 'Computer cartography. Point-in-polygon programs', *Lund Studies in Geography*, C7.

Nordbeck, S. (1970). 'Isarithmic maps and the continuity of reference interval functions', *Geografisk annaler*, **B2**, 92–123.

Nordbeck, S. (1972a). 'Population maps and computerized map construction', *La Revue de Geographie de Montreal*, 1972, pp. 67–76.

Nordbeck, S. (1972b). 'Computer cartography. The mapping system NORMAP. Location models', *Student litteratur*, University of Lund, Lund.

Olsson, A., and Selander, K. (1971). *A Spatial Information System, Dot Maps by Computer* (FRIS C:2). Central Board for Real Estate Data, Gavle, Sweden.

Ottoson, L. (1978a). 'History and Status within Computer Aided Mapping in the National Land Survey of Sweden', Gavle, Sweden.

Ottoson, L. (1974). 'Development of software for numerically controlled draughting for cartographic purposes', Paper presented to the 7th International Conference on Cartography, Madrid, Spain.

Ottoson, L., and Tönnby, I. (1973). 'Numerically controlled draughting of reference grids for cartographic purposes', Stockholm, Sweden, 1973.

Ottoson, L. (1978b). 'Establishment of a high density terrain elevation data base in Sweden', Paper presented to the 9th International Conference on Cartograpy, Maryland, U.S.A.

Sou, K. (1974). *Orter i regional samverkan*. Government Official Reports, Stockholm.

The Computer in Contemporary Cartography
Edited by D. R. F. Taylor
© 1980 John Wiley & Sons Ltd

Chapter 7
Computer-assisted Soil Mapping

STEIN W. BIE

UNDERSTANDING THE SOIL LANDSCAPE

Soil is the loose, uppermost layer of the crust of the earth and is derived from the weathering of rocks and the accumulation of organic matter. It may be *in situ* or transported by wind, water, animals, or humans. Together, the factors of soil formation have created a generally intricate and complex soil pattern. In some soil landscapes some soil properties vary greatly over short distances; others have smaller spatial variability. Other areas may be more uniform or soil properties may show different patterns of variability. Man, through the use of fertilizers and earth-moving equipment, contributes to further modification of these patterns.

In human decision-making on land-use, soil constitutes a factor of uncertainty through the spatial variability of soil properties pertinent to the intended land-use. A *soil map* is information purchased to reduce that uncertainty and thereby hopefully improve decision-making.

Soil science is a new discipline, with its most prominent roots in Russia in the latter half of the nineteenth century. Soil mapping originates from early developments of this science, and has traditionally reflected the largely qualitative approach to the soil landscape. The soil map has been the projection of the complex multivariate soil system on to two geographical dimensions. Until recently, agricultural users constituted the major group of consumers for soil maps. Improvements in agricultural science and technology, and increased contact with more quantitative disciplines (such as civil engineering), are currently generating new sets of specifications for soil maps. In this period of transition the emphasis is shifting from qualitative general-purpose maps to quantitative specific-purpose maps.

During the last 10 years computer-aided methods have won acceptance both as support for traditional soil map-making, and as major components in the development of quantitative methodologies for soil survey. These two aspects reflect two different developments, and it is important to be aware of the distinction. More formally we can express the difference thus: the traditional soil survey procedure has been to classify before interpolation; that is, to map a predetermined legend. The primary task of the survey is to locate boundaries of the mapping units as defined by the legend. Mapping thus focuses on the defining criteria of the mapping units. These criteria constitute a (small) subset of all possible variables. The mapped occurrences are likely to be heterogeneous to a greater or lesser extent. The role of the soil cartographer is limited to giving a useful cartographic expression to the pattern of mapping units sketched by the soil surveyor in the field. An alternative soil mapping procedure is to interpolate before classification. This amounts to estimating the spatial distribution of all relevant soil variables prior either to delineating predetermined classes or using the observations themselves to decide on the classes. When single-property maps are required, the cartographic challenge lies in the selection of a suitable interpolation procedure, and the presentation of the result. When several properties are to be displayed simultaneously, the cartographic problem is compounded by the need to project a multidimensional situation onto two-dimensional paper.

This chapter attempts to review both lines of development, but readers should be aware of the current debate in soil survey organizations on the role of the soil surveyor: Is he/she primarily a *boundary delineator* or a *collector of basic data*? In the first instance, computer-aided soil cartography entails the mechanization of drafting procedures, whilst in the latter case, computer-aided methods of interpolation and classification come within the realm of the soil cartographer. In many ways these two developments are similar to the distinction between automated mapping and computer mapping made by Taylor in Chapter 1.

TYPES OF SOIL MAPS

To the uninitiated, soil is dirt, and the need to classify soil maps may be obscure. A broad division can be made between conventional and computer-derived soil maps, and there are five increasingly complex types of soil maps:

Conventional soil maps
(i) *Manual soil class maps* with predefined legends of soil classes, where the boundaries of mapping units are drawn by the soil surveyor.
(ii) *Manual isoline maps* which are soil maps of a single soil property, the isolines being constructed by the soil surveyor.

Computer-derived soil maps
(iii) Computer-constructed isoline maps where the computer interpolates the isolines from point and line data.
(iv) *Computer-aided soil class maps* which are soil maps with predefined multivariate legends, where the boundaries of mapping units are interpolated by the computer from point data.
(v) *Computer-constructed multivariate soil maps* where both the legend and boundaries of mapping units are calculated from point data.

The computer can, in fact, aid in the construction of all five types but in different ways and to differing degrees. In the subsequent sections an overview of the state of the art is attempted.

COMPUTER-AIDED PRODUCTION TOOLS FOR CONVENTIONAL SOIL MAPS

The manual soil class map and the manual isoline map belong to a category of soil maps for which there is considerable experience with computer-aided methods. The maps are characterized by sets of predefined polygons drawn by the soil surveyor in analog form: the manuscript map.

In the manual cartographic procedures, this manuscript forms the basis for the production of scribe coats and peel coats for colour or black and white production. It is also the basis for the compilation of derived maps, for example soil suitability maps. Essentially, the production process is one of copying the original manuscript, keeping it in an analog form.

Computer-aided procedures were originally brought in to reduce the amount of tedious copying required for the production of the colour plates for coloured maps. Economic motives were important as it was felt that the computer would help reduce costs. It was also hoped that the production of derived (or interpretive) maps could take place from the same computer file used for colour separation.

Towards this end, a number of production tools were introduced, both in the form of hardware (computer machinery) and software (computer programs). Much of the hardware originated from other electronic and cartographic development, but considerable software developments have been unique to soil science and related environmental maps.

Developments have been associated with: digitizing, editing, polygon handling, and plotting.

Some of the features described in the following sections are common to the construction of all types of soil maps.

Digitizing Manuscript Soil Maps

Manual Digitizers

Digitizers used for translating analog soil maps into digital cartographic files suitable for computer handling are the same as those used in other types of computer-aided cartography, as described by Boyle in Chapter 4. They include digitizing tables of the coordinatograph type (Figure 7.1), a similar cross-slide digitizer but with servo follower (Figure 7.2), and the grid type digitizer (Figure 7.3). The digitizing of soil maps poses few requirements beyond those already present for other types of cartography: The menu technique for the entry of text or symbol data is also very common (Figure 7.4).

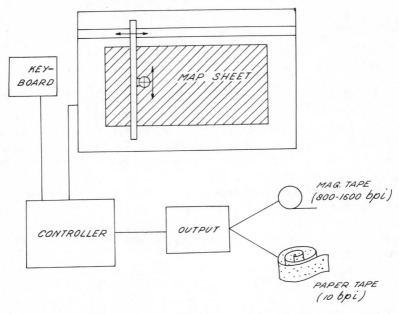

7.1 Coordinatograph cross-slide digitizer. Reproduced by permission of the Centre for Agricultural Publishing and Documentation, Wageningen

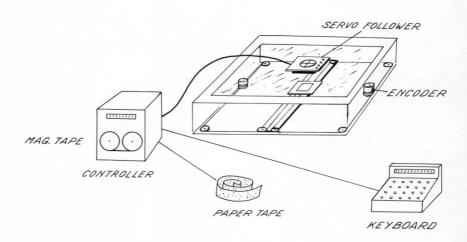

7.2 Cross-slide digitizer with servo follower. Reproduced by permission of the Centre for Agricultural Publishing and Documentation, Wageningen

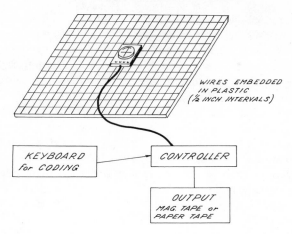

7.3 Mat or grid type digitizer. Reproduced by permission of the Centre for Agricultural Publishing and Documentation, Wageningen

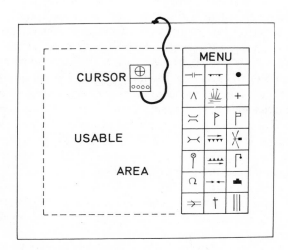

7.4 The menu technique for text and symbol entry. Reproduced by permission of the Centre for Agricultural Publishing and Documentation, Wageningen

However, personnel and organizational situations may differ markedly between relatively small soil survey organizations and those of large topographic institutions embarking on computer-aided cartography. A smaller soil survey institute cannot normally attract new staff to operate the new technology. Existing cartographic staff, with no experience with computer technology, must often be retrained. Complicated digitizing procedures are likely to alienate them.

Ergonomic problems inherent in cartographic digitizing manifest themselves strongly in the digitizing of soil maps. Soil cartographers normally think in terms of areas, which generally constitute the information-carrying unit in soil maps. Knowledge of plausible soil patterns is employed to create an acceptable map image. Digitizing, on the other hand, forces the cartographer to view the map as a selection of lines. It may be advantageous to model digitizing procedures on the previous, more familiar, polygon-oriented pattern, which in turn could lead to reduction in digitizing errors.

A further problem associated with digitizing in general, and no less acute in soil mapping, is the need for feedback during the encoding. Two types of responses are commonly used: a continuous syntactical check on input, and a graphical display of what has been digitized. During the first years of digital soil cartography the *controllers* in Figures 7.1 to 7.3 were merely unintelligent switches and digitizing was *off-line*. This was unacceptable in the long run, and controllers are now minicomputers (or recently, microcomputers) vetting the stream of coordinates and codes being entered. Such *on-line* digitizing now dominates computer-aided soil cartography, and increasingly sophisticated programs are devised for ensuring that data are as correct as possible when stored away for further processing. Graphical feedback can be organized in a number of ways, some using simple marking procedures, others relying on on-line display facilities via the controller.

It is important for the cartographer to review progress in the course of the digitizing operation. A choice may be made between an analog or a digital imprint of the work covered. For soil mapping, the Experimental Cartography Unit (London, England) used a transparent wax paper, on which the scribing cursor left a trace. The Soil Research Institute in Ottawa, Canada, has used an engraving needle attached to the cursor, whereby a scribe coat is produced in the course of digitizing. Groove digitizing, pioneered by the Surveys and Mapping Branch, Department of Energy, Mines and Resources, Ottawa, Canada, has also been done for soil maps. Here, fair drawn soil boundaries are etched into a two-layer film by photochemical means. Coloured wax leaves an imprint of its passage. These methods of recording digitizing progress are, like the marking with a pencil, of analog type and give no indication of the digital information being stored. As rough guides to digitizing progress they nevertheless serve a useful function.

Feedback on the content of the digital cartographic file may take three forms: the off-line check plot; the on-line partial check plot; and the storage cathode ray tube (C.R.T.). The off-line check plot reproduces in analog form the digital file. However, there is usually a considerable time lag between digitizing and check plot, whereby its role as an aid to monitoring digitizing becomes small. The on-line partial check plot provides a running commentary on progress. For this, a combined digitizer–plotter may be used. Figure 7.5 represents a close-up of one model, where the table has both digitizing facilities (of the coordinatograph type) and flatbed plotting capabilities. A transparent overlay on the manuscript soil map provides the writing surface. A major advantage of this technique is the availability of the topographic base map on which the soil boundaries are usually copied. A major problem in digital soil cartography—the fitting of the soil boundaries to features on the analog base map—is therefore solved. C.R.T. storage tubes provide rapid echo of the digital information (see Figure 7.6) but lack the topographic base when this is not in digital form. C.R.T.s are nevertheless the most commonly used display for monitoring digitizing.

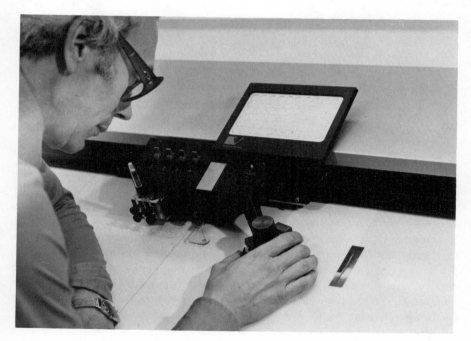

7.5 A combined digitizer–plotter. Reproduced by permission of the Netherlands Soil Survey Institute

Digitizing soil maps also includes the entering of feature codes to denote the name of the mapping units. There are a number of ways of doing this that essentially reflect the degree of sophistication of the computer processing system in associating the feature codes with polygons. A common approach is to declare the feature code on the right and the left of a line about to be digitized. More advanced systems allow the feature codes to be entered subsequent to line digitizing, by computing the relations

7.6 A cathode ray tube used in monitoring digitizing. Reproduced by permission of the Netherlands Soil Survey Institute

between soil boundaries and the polygon name (feature code). This mimics more closely the manual procedure of name placement. Keyboards may be used to enter feature codes, but menu techniques are also common. Here, small fields are associated with programmed instructions which are carried out when activated. Note the difference in design between the menu of Figure 7.4, where the activation is by coordinates of a certain range, and the hard wired menu in Figure 7.5, a pushbutton system. Some early systems employing the former type had the disadvantage of a possible loss of coordinate reference point when the cursor was lifted too high to reach the menu field. It is, in any case, preferable to have a separate cursor for activating such a menu.

There are three basic modes by which soil boundaries may be digitized on a manual digitizer. Figure 7.7 describes the point mode, time mode, and incremental mode. Point mode digitizing tends to lead to smaller data files. However, the larger files from time to time or incremental digitizing may be 'weeded' to remove those points that do not add significantly to the positioning of the line. Angle and between-point-distance are commonly incorporated in 'weeding' algorithms.

Digitized soil maps, like all other digital maps, contain errors. Such errors not only lower the quality of the map image, but can also make the digital cartographic file almost unusable. In common with most polygon maps, a digital soil map must be completely free of logical errors to be usable.

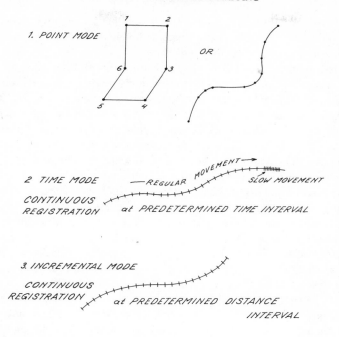

7.7 Basic digitizing modes. Reproduced by permission of the Centre for Agricultural Publishing and Documentation, Wageningen

Fatal errors are those concerned with the logic of the polygon nets. Undershoot and overshoot of lines represent the most common errors in digital soil maps. Undershoot and overshoot prevent the polygons from being logically closed; this is not necessarily always visibly apparent. Polygons cannot be formed and map construction will fail. There are various ways of correcting these problems, but in terms of digitizing strategy there are also ways of avoiding them.

The most direct approach is to digitize the intersections of boundary lines separately (as *nodes*) and to define all line segments as originating and terminating in predefined nodes. Islands may be defined by one or two nodes. This procedure prevents undershoot and overshoot by definition, but does require more digitizing effort. An alternative approach is to define the first and the last digitized point of a line as nodes, and to use software to search for already existing nodes to avoid duplication.

Automated Digitizing

Manual digitizing is a tiresome and repetitive process. It is increasingly realized that large-scale computer-aided cartography for soil maps will only be successful if more efficient methods are found for creating the digital data file required. Unfortunately, automated digitizers are still in their infancy, although as Boyle (Chapter 4) indicates, considerable progress has recently been made.

Automated digitizers are either of the scanner or the line follower type with scanners being more promising. Scanners split the image into a large number of small cells (raster image), whilst line followers lock onto lines and deliver the coordinates on normal (vector) format. The critical elements in the design of raster scanners are their resolution, data compaction, and raster-to-vector conversion. Line followers face decision problems at line intersections, and when arriving at dead ends. Both types of devices are usually unable to separate soil manuscript lines from background base map detail, which means that an overlay, with soil boundaries only, has to be prepared for digitizing. When this has to be a separate operation, the saving compared to manual digitizing may be small. Soil survey organizations possessing the films for colour separations of the soil boundaries may find these suitable for automated digitizing.

Boyle has indicated the possibilities for scanners but in 1979 there was only one soil survey organization that routinely employ automated scanning for the production of soil maps and derived maps. The experiences of the Soil Conservation Service (U.S. Department of Agriculture, Hyattsville, Maryland, U.S.A.) however, suggest that given sufficient resources both for computer systems and procedural reorganization automated digitizing may prove successful. The technical feasibility of scanning has been further demonstrated both by the early work of Canada Geographic Information System (Lands Directorate, Environment Canada, Ottawa, Canada) and current activities by the Geography Program (U.S. Geological Survey, Reston, Virginia, U.S.A.) where service bureaus have been employed for automated digitizing.

The quality of automated digitizing is at the moment such that considerable editing is required to obtain a file fit for polygon construction. The actual time required for scanning (or line following), with subsequent processing to obtain vector files, varies, but is commonly of the order of $\frac{1}{2}$–1 h for a map sheet. Editing times, using interactive facilities, may run from 10 to 30 h, depending upon the complexity of the soil map and the quality expected. This compares to 50–100 h commonly recorded for manual digitizers of comparative maps, including editing. Depending upon the rate of amortization of facilities used, salary levels of operators, and overheads, digitizing and editing costs in Western Europe and North America were ten to U.S. $25 in 1978. Service bureaus in Europe quoted, in 1978, around U.S. $2000 for the digitizing of soil maps 40 × 60 cm. The client must expect to do some editing of the digitized product, say U.S. $500 worth. Against this must be seen the costs of hardware and software which at the moment are about ten to twenty times those of manual digitizers. There is every hope that technological progress will alter this picture and thereby remove the number one bottleneck in automated soil map production. Boyle has clearly indicated that most problems are now being solved and that scanners will become increasingly more common.

Editing

The editing of digital soil maps represents the updating of the digital file. Editing may serve three basic functions:

the correction of errors generated during digitizing;
alterations in the file due to new information;
changes associated with plotting requirements.

Until recently the common way of correcting soil maps was to work in off-line mode. Errors found on a check plot were corrected by removing offending segments (frequently by removing punch cards) and by entering new segments from the digitizing table. Similar procedures were used for feature codes. This procedure was very cumbersome.

Interactive editing systems are now replacing redigitizing as a means of updating the file. Figure 7.6 shows a commonly used editing station, with a C.R.T., a keyboard, and a small digitizing table (graphic tablet) often including a limited menu. Here, the editing station operates on-line to a minicomputer. Offending lines or feature codes are pointed to on the screen using a cursor, and editing commands issued for the removal, transposition, or other alteration of the file, whereupon the corrected result is displayed for acceptance. Useful editing commands are described in the section entitled 'Typical Software Requirements'.

It is recognized that interactive editing, although a major improvement over the older type of line editing, is still a time-consuming task, for data generated both by manual and automated digitizers. Manufacturers of interactive systems now tend to do some editing in simulated batch mode (or absentee mode), where major checking and error identification are carried out without user interference. Recent work by the Norwegian Computing Centre in Oslo on similar maps suggests that three-quarters of all errors may be removed by software either residing in the minicomputer or by host computer processing.

Polygon Handling

The soil polygons are the information-carrying units of the soil map. With the exception of polygons along the map edges, all soil polygons share sides with other polygons. There have been a multitude of ways of expressing these relations in computer-aided soil cartography.

Early developments included the complete digitizing of each polygon, thus generating duplication of polygon sides. This created many problems of overlap, misfit, and double digitizing.

The common way now of organizing polygon data files is to define the polygon through a series of pointers in the data file to the polygon sides. In the early days, the entry of lines in a constant direction (for example, clockwise) coupled to a right/left margin indication of feature codes at the time of manual digitizing could be used to create the polygon. Interactive methods were also in use, whereby the pointers for each polygon were constructed by pointing to polygon sides, one by one, on an interactive C.R.T. screen. Current methods execute polygon contruction in absentee mode, solely based on coordinate data. This is now the only sensible method for routine production of soil maps.

Once the polygons have been constructed, the name (feature code) of each polygon must be associated with it. This is achieved by creating a pointer to the polygon file. Earlier this was done interactively by pointing on a C.R.T. Recent development allows for automatic polygon labelling. As long as the polygon feature code is inserted somewhere inside the polygon, a location algorithm may be used to erect the pointer.

The production of colour separations for colour printing can be carried out once all polygons have been sucessfully constructed and labelled. The logic of colour separation by computer is identical to the manual method. Polygons with particular feature codes are assigned colour and percentage (or screen) in the colour scheme. For each percentage (or screen) of each colour a plot must be made. To create a plot suitable for the now commonly used *open window* lithography, only those lines may remain that border areas to be peeled from the peel coat. External lines must be removed, and internal lines (common lines) which are superfluous must be similarly suppressed.

A common way of achieving this is to associate with each line a pointer which may be set to visible or invisible in the digital file. The colour separation is done by a program that sets this pointer to the correct value, and allows the resulting visible lines to be copied onto a separate subfile for plotting later. Another program may be used to return all pointers to visible, ready for the next colour separation.

Colour separation executed this way constitutes a very rapid method for preparing the plots for the lithographic process. Programs to execute colour separation may be executed on minicomputers (in absentee mode) or on large computers in batch mode. The preparation of a colour soil map may commonly involve up to thirty to forty plots.

Derived (interpretive) maps, for example soil suitability, may be created using the same logic and the same computer programs as for colour separations. However, only one plot is necessary. Internal (common) lines must be removed. As with colour separations, the common way of achieving this is by renaming the polygons using a correspondence list, where soil feature codes are translated into, for example, soil suitability codes. Then the pointers for visibility are set to visible or invisible. Figures 7.8 to 7.11 illustrate the use that may be made of a soil map (Figure 7.8) and a ground water map (Figure 7.9) combined to form suitability maps for arable agriculture (Figure 7.10) and recreation (Figure 7.11) of an area in the Netherlands.

Since users of soil maps may well be more interested in derived maps than the soil maps themselves, the ability to sort the original feature codes and remove unwanted common lines are essential parts of a successful computer-aided system.

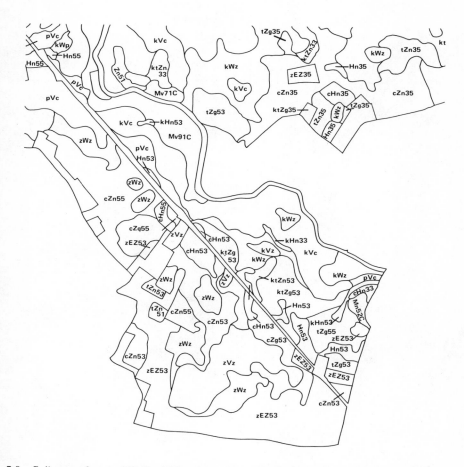

7.8 Soil map of part of Holland. Reproduced by permission of the Netherlands Soil Survey Institute

7.9 Ground water map of area shown in Figure 7.8. Reproduced by permission of the Netherlands Soil Survey Institute

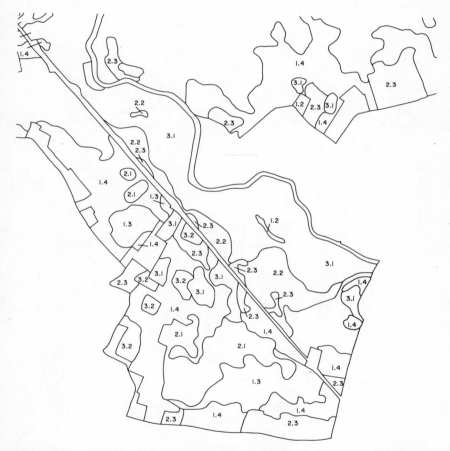

7.10 Suitability of area for arable agriculture. Reproduced by permission of the Netherlands Soil Survey Institute

7.11 Suitability of area for recreation. Reproduced by permission of the Netherlands Soil Survey Institute

An extension to the cartographic techniques outlined in this section is the ability to associate further information with the polygons; that is to link a property file to the cartographic file. This file may contain descriptions of commonly occurring kinds of soil and their properties in the form of attribute value pairs. Very few computer-aided cartographic systems currently offer this facility. When they do, full logical and arithmetic search facilities (i.e. AND, OR, NOT, $=$, $>$, $<$, etc.) should be available. Such property files may be associated through pointers to polygon or feature codes. An elegant addition is to allow selected values from the attribute value file to be displayed with the cartographic information.

Plotting

A common complaint levied against computer-drawn soil maps has been associated with so-called poor 'cartographic quality'. This refers to the line quality of the finished product. (In other respects computer-drawn soil maps may be expected to be better, for example, with perfect colour separation) Two factors are particularly associated with this:

    smoothing of lines;
    technical specifications of the plotter

Whatever mode of digitizing is chosen (see Figure 7.7) the coordinate stream residing in the file must be converted to lines. In the simplest case this is done by straight line segments, which will yield a jagged outline. Although it seems unlikely that the predictive value of the soil map is impaired by this, its aesthetic value is. Smoothing of the lines is therefore desirable, which involves the application of some geometric function (normally a polynomial) to the digitized points between two nodes. Figures 7.8 to 7.11 have been smoothed using a spline function technique. This does not remove the limitations inherent in the mechanical design of the plotter used. Every plotter has a minimum step size, between which a straight line (or, in some advanced systems, an arc) is drawn. Smaller survey organizations have normally not had the financial resources for the purchase of plotters with small step size. They currently cost in the order of U.S. $100,000–200,000. New advances in plotters with outputs on microfilm, or even more accurately, laser plotters, may reduce the costs of high quality plotters. Boyle considers this fully in Chapter 4.

The use of an inferior plotter should in no way distract from the value of a computer-aided system. Even slightly wriggly lines convey important soil information. The choice of drum plotters, simple flatbed plotters, electrostatic plotters or microfilm systems to convey digital soil class maps is mainly a matter of financial resources and if these resources are available very high-quality results can be obtained.

Manual Isoline Maps

Isoline maps constructed by hand constitute a special case of polygon maps. Isoline maps are commonly made to show, for example, the depth to bedrock, depth of sand, salinity and other chemical properties that are continuous variables.

So far such soil maps have rarely been digitized for cartographic purposes. When they are, the techniques employed are similar to those described for the manual class map. However, it is useful to be able to attach the value of the isoline to the line itself, in addition to creating a feature code for the between-isoline areas. They are edited and plotted as soil class maps, and colour separations may be made.

The main reason for digitizing isoline maps would seem to be the creation of digital information for non-cartographic usage, or as input for methods of computer-constructed soil maps discussed in later sections.

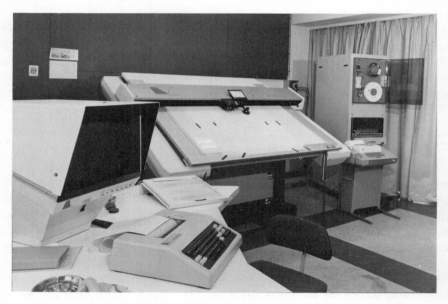

7.12 Typical hardware associated with a computer-assisted soils mapping system. Reproduced by permission of the Netherlands Soil Survey Institute

Typical Hardware Requirements

Current systems with the capabilities described in the above sections include one or more digitizing tables, a cathode ray tube, a plotter, a magnetic tape unit, exchangeable disk drive, and a minicomputer together with alphanumeric terminals. Figure 7.12 illustrates one such system with a combined digitizer/plotter (the disk unit is hidden). It is common experience that soil maps, with their colour separations, require much disk space, and at least 24 megabytes. Minicomputers are normally 16 bit machines, with from 64K core. Systems should have the ability to maintain several digitizing stations, plotters, or C.R.Ts simultaneously. This normally requires more core and disk space. Magnetic tape units should be industry compatible.

Typical Software Requirements

Users frequently experience the need to program their own modules or to modify the software supplied by maufacturers of the hardware components purchased. It is therefore important that software agreements allow such modification or additions, and that a high-level language is available for programming. Systems supporting FORTRAN have enjoyed a distinct advantage in this respect.

Whilst not claiming to be exhaustive, the following list may provide a guideline to functions found useful for computer-aided data handling of conventional soil maps.

Function	Comment
DEFINE ORIGIN	Defines reference point on map in terms of map and table coordinates
DEFINE SCALE	Establishes conversion between map distances and distances on the ground
CALIBRATE X-AXIS } CALIBRATE Y-AXIS }	Allow independent adjustment of map axes to compensate for linear distortions in manuscript maps
RECTIFY MAP	Allows for non-linear adjustment of the map by comparing digitized control points with known coordinates
CREATE NODE	Locates future intersection prior to creation of the lines
CREATE LINE	Enables line to be digitized between nodes
INSERT TEXT	Allows the placing of feature codes (upper and lower case)
INSERT PROPERTY	Allows attribute value pairs (to be defined) to be associated with cartographic feature
DELETE NODE	Deletes node and all lines associated with it
DELETE LINE	Deletes line but leaves nodes
EDIT TEXT	Enables the removal or changing of text
BREAK LINE	Severs a line and inserts a node in breaking point, attaching line segments
CONSTRUCT POLYGONS	Attempts polygon construction, denotes errors by messages and error map
NAME POLYGONS	Associates feature codes with their respective polygons and denotes errors by messages and error map
EXECUTE COLOUR	Creates subfile of named polygons and deletes external

SEPARATION	lines and common lines. Feature also to be used for construction of interpretive maps
UNBLANK FILE	Sets pointers of master file back to visible
ZOOM UP/DOWN/ RATIO/WINDOW	Enlarges (or reduces) size of image on C.R.T. by explicit or implicit ratios
PLOT	Plots file with given scale, speed, pen, window

In addition to the above commands, which specifically relate to the construction of soil maps, a large number of 'normal' system commands are required for the efficient functioning of a computer-aided soil cartography system.

COMPUTER CONSTRUCTED ISOLINE MAPS—AUTOMATIC INTERPOLATION

We can never hope to examine all points making up the soil landscape, so if we are to make statements about the soil at unvisited sites, we must interpolate from the known values.

When the attribute considered is a continuous variable (depth to ground water, concentration of a chemical property) there are established techniques for interpolation. All algorithms that allow such estimations use mathematical models that may or may not happen to bear relation to the physical model responsible for the spatial distribution of values.

There are four main models available for interpolation:

grid methods, involving the placement of a netlike structure on the data, and the use of estimated values at grid points for interpolation;
triangulation, using triangles drawn between points for estimation;
polynomials, fitting polynomial functions to series of observation points;
kriging, an advanced moving-averages estimation technique utilizing the estimated pattern of the spatial variability of the property under consideration.

Although contouring techniques are well established in other environmental sciences, there are no studies known to this author with a critical evaluation of alternative procedures available for soils data. Contouring is usually carried out with a contouring package available at a particular computing centre. General-purpose contouring packages are usually implemented on large computers only, and some are very costly. Potential users of contouring packages may obtain useful suites of programs from academic or research institutions at nominal prices (for example, SYMAP from Harvard University, Cambridge, Massachusetts, U.S.A.; or SURFACE II, from Kansas Geological Survey, Lawrence, Kansas, U.S.A.)

Since so little practical experience is available for soils data, and soil scientists usually are small users of contouring packages, one may have to rely on what is available. The following criteria should be met when using a contouring package:

- the programs accept unevenly spaced data points;
- the user can influence the search radius of the algorithm;
- the user can choose grid size (if appropriate);
- that discontinuities (faultlines, excavations, water divides) can be accommodated;
- varied possibilities for graphical output are available (line printer, plotter, C.R.T.)

Figure 7.13 illustrates an isoline map produced on a lineprinter to show topographic variation, and Figure 7.14 is an isoline map of ground water level where several barriers have been introduced to account for watersheds (see also Figures 7.16 and 7.17 for further use of the information contained in Figures 7.13 and 7.14).

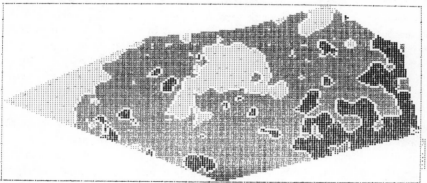

7.13   Isoline map of topographical variation produced on a line printer. Reproduced by permission of the Netherlands Soil Survey Institute

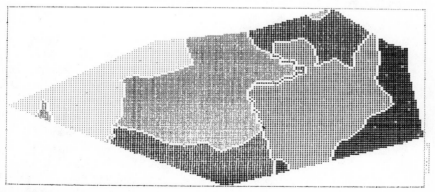

7.14   Isoline map of ground water using barriers to alter interpolation. Reproduced by permission of the Netherlands Soil Survey Institute

COMPUTER-AIDED HANDLING OF MULTIVARIATE MAPS WITH FIXED LEGENDS

There is an inherent difficulty in trying to project the spatial variability of many soil properties simultaneously on two-dimensional paper. The common solution has been to classify first, and then to delineate the classified areas. When the delineation is done manually, we get the manual soil class maps discussed in the first section. However, we may also attempt to delineate classified points by computer-aided methods. We may distinguish between two inherently different approaches: proximal mapping and polygon overlay techniques.

Proximal Mapping

If we have been able to classify the soil encountered at an observation point; that is, allocated it to a class in an established soil classification, we may proceed to delineate boundaries between different soil classes using a mathematical model. Since soil classes are not continuous variables, isoline techniques are not appropriate. The most commonly known model for the construction of such *proximal* maps involves the use of so-called Thiessen polygons. This is a geometric construction technique whereby boundaries are constructed midway between observation points belonging to different soil classes. It should be obvious that this is a primitive model for displaying multivariate variability. There is little published work in soil science using proximal mapping, but it is known that experiments have yielded interesting results, particularly where the geographical spread of the variables is even. There are similar geometric methods suggested for the construction of soil maps. Figure 7.15 reproduces a map of soil parent material achieved by expanding print cells from known observation points, and outputting the map on an electrostatic printer/plotter.

Polygon Overlay Technique

If we had separate maps of each soil variable to be considered in an area, we could overlay transparent copies of the maps to produce an optical mixing of the information: colours may be useful to create distinct patterns. Computers may also be used for this purpose, and calculations made where individual polygons intersect or join each other. This technique was originally developed for the Canada Geographic Information System in Ottawa. Although technically competent, it illustrated the problem of fragmentation when two or more nearly identical polygons were overlaid. Small strips of new combinations arose along the edges of the 'not quite matching' polygons. The Bureau of Land Management, Denver, Colorado, U.S.A. is currently designing a large system with modules intended to overcome this problem. Overlaying polygons is a very computer-demanding exercise, and raises many problems of both conceptual and cartographic nature. It has not yet found general acceptance.

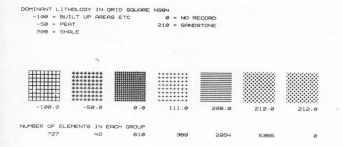

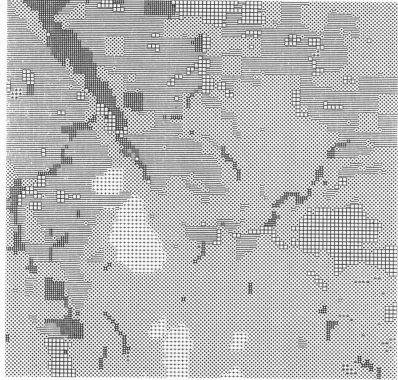

7.15 Map of soil parent material produced on an electrostatic printer/plotter. Reproduced by permission of the Macaulay Institute for Soil Research

An alternative technique has recently been suggested which may also involve a change in the normal sequence of classification before interpolation. It has been suggested that each soil variable, whether continuous or not, may be interpolated separately prior to classification on the basis of a set of variables for each observation point. If an area is divided into small cells (raster format), a class allocation algorithm may be devised so that each cell is allocated to a class on the basis of the values each cell has for various attributes. Figure 7.16 shows the soil map of the area covered by Figures 7.13 and 7.14, but split into cells. An allocation algorithm was devised to calculate drainage charges to farmers based on the difference between altitude and ground water level seen in the light of soil conditions (Figure 7.17). Figure 7.18 presents the synthesized water drainage levy map constructed in this way. Unlike manual soil class maps and proximal maps, the spatial pattern of *each* variable has been allowed to influence the final map image. This allowance for spatial variability separates this map from the other types where classification only considers variability in property space.

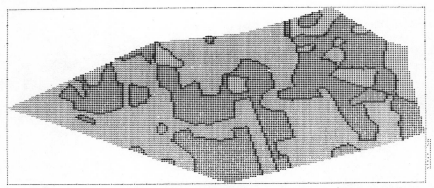

7.16 Soil map on a cell-by-cell basis. Reproduced by permission of the Netherlands Soil Survey Institute

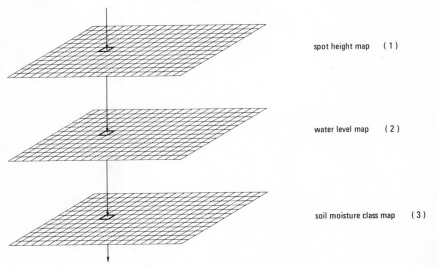

7.17 Allocation algorithm for drainage charges. Reproduced by permission of the Netherlands Soil Survey Institute from Bie *et. al.*, *Neth. J. Agric. Sci*, **26**, 278–287(1978)

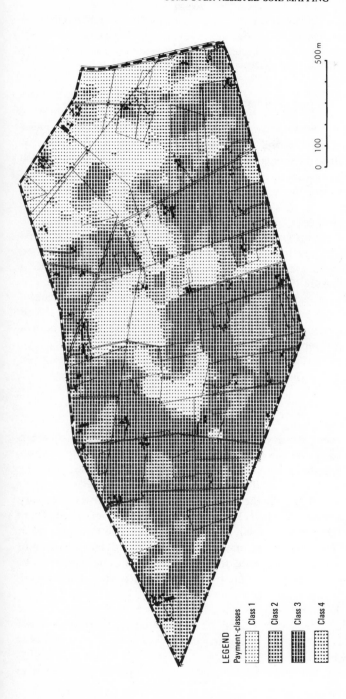

7.18 Resulting drainage payment classification map. Reproduced by permission of the Netherlands Soil Survey Institute from Bie et al., Neth. J. Agric. Sci, **26**, 278–287(1978)

Similar techniques are currently used in the evaluation of soil behaviour in landslides and earthquakes (by U.S. Geological Survey, Reston, Virginia and Menlo Park, California) and evaluation of soil (overburden) properties for strip mining and restoration (Geological Survey of Canada, Calgary, Alberta).

COMPUTER-CONSTRUCTED MULTIVARIATE MAPS

Whereas the maps in the previous section were constructed by allocation to a given (fixed) legend, there are also possibilities for constructing both legend and map automatically from multivariate data. The method, originally suggested by De Gruijter and Bie (1975) and supported by a variety of multicoloured maps in De Gruijter (1977), involves using the values for soil properties at the observation points to erect a soil classification using numerical taxonomy. Clustering techniques are employed to obtain optimal classification in property space, based on the observations available. Once the classification is established, an allocation employed to obtain optimal classification is established, an allocation algorithm is devised whereby an unclassified unit can be allocated to an existing class through a multivariate distance measure (for example, Euclidean distance). Each variable is then mapped separately, in raster format and using isoline techniques. A stack of values for one cell can then be used to allocate that cell to one of the classes. This technique is an attempt to optimize the projection of a multivariate set of data both in property space and in geographical space.

Alternative techniques involve the contouring of the multivariate measure from property space; Webster and Burrough (1972) demonstrated this approach.

CONCLUSION

Computer-aided techniques for soil mapping are of recent date, and have as yet had only minor impact on soil survey organizations.

Current experience indicates that the high cost of digitizing manual soil maps by hand makes the mechanization of the drafting process unattractive to many organizations. It is also uncertain whether manual digitizing can offer the necessary job satisfaction in the long run. We must hope that progress in automated digitizing can make a significant contribution to reducing costs, and as Boyle has indicated in Chapter 4, substantial progress is being made. Although the automated drafting of soil maps is useful it is the ability to create derived (interpretive) soil maps from the basic soil information which holds more interest for the soil scientist. This allows soil science to come closer to the user of soil information.

Recent suggestions of improving soil classification and interpolation of soil properties in geographical space by using statistical methods currently reflect research interest, and it remains to be seen whether this will have an operational impact on soil cartography.

However, it does seem that the availability of soil geographical information in digital form, as computer files, opens a whole new role for soil maps. The ability to store the raw data efficiently in such files, to merge them and derive cartographic data using advanced statistical techniques, point to the development of integrated soil information systems. Here, soil maps will continue to provide a major form of display of soil data. Manually compiled soil class maps are, however, likely to be superseded by computer-derived maps optimizing data presentation specific to a user's needs in the near future. The present-day soil maps convey too limited an amount of the total information collected by the soil surveyor for its preparation. The current development trends indicate that both the scientific content and user-acceptability of soil maps may be significantly enhanced by computer-aided methods for soil cartography.

Acknowledgement

The support of Netherlands Soil Survey Institute, Wageningen, the Netherlands, is gratefully acknowledged.

REFERENCES

Bie, S. W. (ed) (1975). *Soil information systems*, p. 87. PUDOC, Wageningen.
Bie, S. W., van Holst, A. F., and Stricker, J. M. N. (1978). 'A computer-aided method for the estimation of water drainage charges—a Dutch case study, *Netherlands Journal of Agriculture Science*, **26**, 278–287.
De Gruijter, J. J. (1977). *Numerical Classification of Soils and its Application to Survey.* Agricultural Research Reprint 885, PUDOC.
De Gruijter, J. J., and Bie, S. W. (1975). 'A discrete approach to automated mapping of multivariate systems', *Proceedings of the International Cartographic Association Commission*, **III**, 17–28.
Ragg, J. M. (1977). 'The recording and organization of soil field data for computer areal mapping', *Geoderman*, **19**, 81–89.
Webster, R., and Burrough, P. A. (1972). 'Computer-based soil mapping of small areas from sample data', *Journal of Soil Science*, **23**, 210–234.

The Computer in Contemporary Cartography
Edited by D. R. F. Taylor
© 1980 John Wiley & Sons Ltd

Chapter 8
Geological Mapping by Computer

CHRISTOPHER GOLD

INTRODUCTION

Davis (1973), p. 298 states: 'Maps are as important to earth scientists as the conventions for scales and notes are to the musician, for they are compact and efficient means of expressing relationships and details. Although maps are a familiar part of every geologist's training and work, surprisingly little thought has gone into the mechanics and philosophy of geologic map making.' He then states that most of the work in this field is done by geographers, and gives various examples. One of the difficulties with this approach of merely accepting rather than developing techniques is that many of the wide variety of map types that geologists use are uncommon in other disciplines.

Maps are displays of items of information that have both *spatial* and *non-spatial* components—without both of these it is not possible to produce a true map. Data that are not in suitable form must be transformed in order to make map production possible. Statistical data, for example, do not usually possess information about the location of each data point in space, but scatter plots or contour plots are frequently generated from data of this type, by arbitrarily considering two variables to represent 'east' and 'north' and then either plotting data point locations or contouring the value of some other associated variable. A point to note is that, given points within a scatter diagram, it is often desirable to contour their density. This density is a continuous variable expressing the number of points per unit area. Unfortunately, the resulting surface varies greatly with the sampling area chosen. This transformation from point location to density has been studied by various geologists, in particular Ramsden (1975, and in press).

Perhaps surprisingly, one-dimensional maps are commonly used by geologists: the geological section is the basic tool of much field work. The direction is usually vertical and the mapped variable may be either categorical, as in stratigraphic sections showing the rock types and formations present in some locality, or continuous, as with electric log measurements of resistivity, etc., down a drill hole. Much geological work consists of taking these one-dimensional maps and correlating between them to produce two-dimensional cross-sections (Figure 8.1A) which imply, conceptually if not diagrammatically, three-dimensional models (Figure 8.1B).

Alternatively, the three-dimensional model may be constructed by generating contour maps of each geological contact in turn.

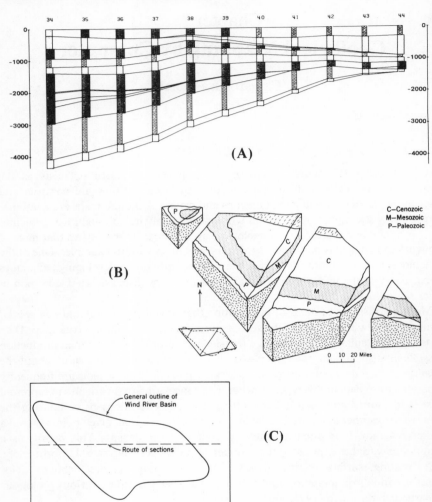

8.1 Wind River Basin, Wyoming, stratigraphic cross-section and block diagram (Harbrough and Merriam, 1968; reproduced by permission of John Wiley & Sons, Inc.)

If only spatial information is present in a data set it may become necessary to eliminate one dimension by treating it as non-spatial for display purposes. An example is the case of a topographic map where east, north and vertical measurements are obtained for points on the atmosphere–lithosphere contact. In this case the elevation information is treated as non-spatial and displayed in various ways on the two-dimensional map sheet—perhaps by plotting spot elevations, colouring elevation zones or some other technique. It is, of course, possible to choose a different pair of map variables and draw a cross-section to the topographic surface, usually with the y-axis of the map representing the elevation.

Geologists basically study the earth, a three-dimensional object, and are frequently concerned with the variation of one or many parameters within three-dimensional space. This differs markedly from political scientists, for example, who rarely consider the vertical variation in voting preferences! (Both groups, however, would frequently like to be able to represent time as another spatial dimension.) This underlying concern with what is happening in the third, undisplayed, spatial dimension is of great importance in some, if not all, branches of geology. Considerable use is made of mapped data containing no intrinsic spatial information; for example, in examining how one measured chemical component in a set of rock samples varies with changes in another component. Most maps make use of data with only two spatial variables—examples of these are maps of the earth's magnetic field, concentration of some economic mineral or element found in an area, or the type of rock to be found in a particular geographic region. This last is perhaps what most people think of as a geological map. A classical example of this is shown in Figure 8.2, the first true geological map, drawn by William Smith and published in 1815 and showing the major rock units in England and Wales. Nevertheless, a little thought will suggest that all of these examples are two-dimensional only because of the inability to collect samples far from the surface of the earth, not from limitations in the display technique. Indeed, many maps are intended to facilitate the intuitive estimation of variation in the third dimension, based on professional knowledge of the behaviour of the parameters displayed (see the discussion in Gold and Kilby, 1978). This is especially true for maps displaying structure and lithology, where structural boundaries and plotted bedding orientations are designed to enhance the three-dimensional model perceived by the compiling geologist.

Maps that are usually amenable to consideration as two-dimensional include many *geophysical* maps (geomagnetic, gravitational, etc.), *geochemical* maps (plotting the concentration of some element measured in samples of soil, rock, or water obtained near the earth–air or earth–water contact) and *facies* maps (partitioning some geological contact into zones with relatively homogeneous properties). *Orientation* and *density* maps will usually be two-dimensional as well. Depending on the scale and the objectives of the study, *'geological'* or, more properly, *'lithological'* maps may be considered to be two-dimensional, but as previously noted the third dimension may be indicated either by bedding orientation symbols or by the form of the geological contact, especially as it relates to the topographic surface. *Topographic* maps, *structural contour* maps, *isopach* maps, and other derivatives are of course three-dimensional in input, even if the vertical dimension must be treated as a mapped variable for display. These map types, along with related discussions of *data capture* and *data base management*, will be described under separate headings later in this chapter.

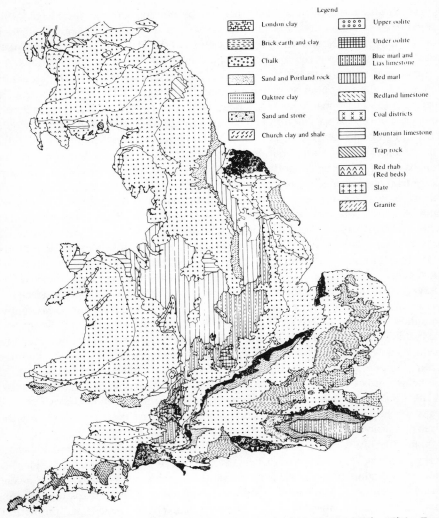

8.2 Part of William Smith's geological map of Britain, 1815. Figure 1.7 (p. 10) in *Earth History and Plate Tectonics*, 2nd Edn, by Carl K. Seifert and Leslie A. Sirkin. Copyright © 1979 by Carl K. Seifert and Leslie A. Sirkin. Reproduced by permission of Harper and Row, Publishers, Inc

Perhaps it is worth making a comment here on the role of maps in geology. They are certainly a product—most geologists have produced maps. As well as output they are also input—all geologists have read maps; many kinds of maps. Rather than being attractive products to pass on to the end-user, most maps are an essential, and primary, means of communication between geologists, and most are not expected to be meaningful even to those geologists not concerned with the particular topic under consideration. Since relatively few maps are intended for immediate comprehension by the general public, a high level of generalization is not usually desirable. Hopefully this explains, and even excuses, the great amount of detail found in many geological maps.

Having mentioned the role of maps in geology, we come to the main purpose of this chapter—to discuss the role of the computer in producing geological maps, and in particular how the advent of the computer has affected geological map production. The main details of these changes are described under each map type. In general, the first change produced by computer availability was the dramatic increase in the quantity of data points generated, with the consequent necessity to display them. This is still the driving force behind map automation. The first mapping techniques developed in response were contouring algorithms, and these are still the leading mapping functions used. It is becoming evident, however, that some of the earlier and more inefficient approaches to contouring are beginning to show signs of wear under the ever-increasing load of generated data.

Apart from contour mapping, the increasing volume of data has mainly generated programs for individual data point display ('posting') and statistical and other data transformations. Of course, once it became easy to generate many maps from one data set, many maps were requested. One offshoot of this is that it is becoming more common to standardize the acquisition of field data by using coding forms, thus facilitating data entry for computer generation of base maps that rapidly display the raw data for subsequent interpretation. It is now not unknown for the weekly mail plane to arrive at the bush camp to collect the week's coding forms and deliver the latest base maps.

The primary effect of computers has thus been to facilitate the production of ever-increasing numbers of simple maps. In this way the geologist may more rapidly reach the stage of data interpretation, where his comprehension of geological processes may help him to make meaningful decisions from the displayed information. Nevertheless much research is in progress in other directions, often with the intention of computerizing or standardizing even more of the routine labour of the practising geologist.

GEOPHYSICAL MAPS

Geophysical mapping involves the display of a wide variety of physical parameters, for example the earth's gravitational field, the earth's magnetic field (both strength and orientation) and earthquake data. Also mapped, on a more local scale, are resistivity and self potential data and, even more commonly, seismic data for oil exploration. A wide variety of less common data types are also used.

Most of these measurement techniques have some features in common—large quantities of data are accumulated by automatic means, usually along flight paths or traverse lines. The data processing is frequently sophisticated and beyond the scope of this study. Nevertheless difficulties exist in the management and display of the large quantities of data of this type that are relevant here. The key point, however, is that the display of large quantities of data necessitates the use of computer-based mapping systems.

While many types of geophysical measurement are possible, examples will be given for geomagnetic surveys (examining the earth's magnetic field either for regional trends or local anomalies); gravity surveys (as above, but for the gravitational field); seismic 'profiling' to identify geological contacts by monitoring the reflection of shock waves; and regional seismic studies to determine earthquake locations.

Dawson and Newitt (1977) produced a regional map of the earth's magnetic field for Canada, based on 51,133 data points (Figure 8.3A) for which the earth's field had been obtained in three perpendicular directions. These data were obtained from an on-going data collection process from 1900 to the present although, since all three components of the earth's field had to be measured at each location, the data points actually used were acquired between 1955 and 1974. Observations were obtained either from ground stations or, more recently and for the bulk of the data, from aerial surveys.

Since this was a regional survey, concerned with field fluctuations of the order of 1000 km in wavelength, it was estimated that a sixth-degree polynomial fit in each quarter of the map area would display such features. In addition, since the earth's field varies with time, a cubic time component was used to adjust for data acquired at different dates. The resulting charts were plotted for estimated 1975 values and a correction applied to correct to sea-level elevation. These charts displayed D (magnetic declination), H (horizontal intensity), Z (vertical intensity), I (magnetic inclination) and F (total intensity). These parameters (which are merely different ways of displaying the orientation and intensity of the field calculated using the orthogonal x, y, and z vectors), were plotted using the regression coefficients obtained independently for the x, y, and z components. Figure 8.3B shows the map for magnetic declination.

Kearey (1977) described a gravity survey of the Central Labrador Trough, Northern Quebec. Average sample spacing was 5 km, with some detailed traverses having a spacing of 1.5 km or less. Regional and residual gravity fields were estimated. The residual field was then related to the lithologic variation within the Trough.

Howells and Mackay (1977) performed a seismic reflection survey of Miramichi Bay, New Brunswick. A more common type of survey uses the refraction technique, which determines the propagation velocity of seismic energy through rocks and surficial sediments. Contrasting lithologies may be correlated with certain velocities, or velocity ranges, provided borehole control is available. This approach is the one primarily used in oil exploration. In the Miramichi study, however, the water depth and available equipment made a reflection survey more applicable.

The objective of the survey was to determine surficial sediment types and

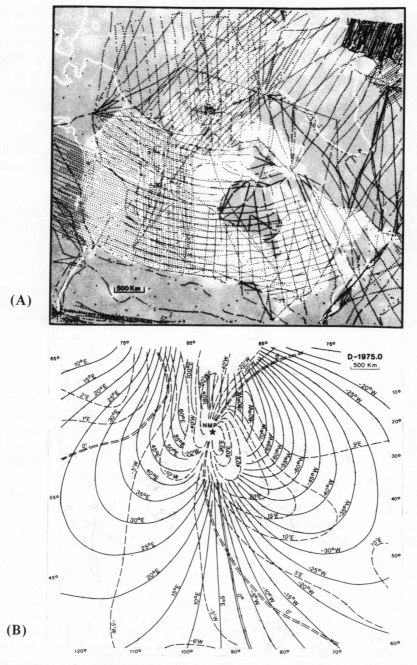

8.3 Distribution of magnetic survey data for Canada 1955–73 and the resulting estimated magnetic declination for 1975. (Broken lines show annual change in minutes of arc (Dawson and Newitt, 1977).) Reproduced by permission of the National Research Council of Canada from the *Canadian Journal of Earth Sciences*, **14**, 477–487 (1977)

thicknesses in the area, and over 300 line kilometres of seismic profiling was obtained. The equipment used consisted of a launch-mounted echo-sounder receiver with the 'boomer' sound generator towed behind on a catamaran. Position fixing for the survey was accomplished using a microwave system with three pairs of land-based slave stations. The overall position-fixing repeatability was about 10 m. The inner part of the bay was less than 7 m deep; the outer part sloping, to 15 m.

In the presence of soft bottom muds a good reflection was received from the harder underlying material. Where the sediments were harder penetration was not achieved, but sea floor roughness was used to distinguish relatively flat sands from the rougher exposed bedrock. This interpretation was aided by several shallow boreholes. On the basis of these measurements various conclusions were made about the geological history of the area.

Basham *et al.* (1977) examined the last data type we will consider under the heading of geophysics—the incidence of earthquakes. From the catalogue of Canadian earthquakes, they extracted all those with epicentres north of 59 degrees of latitude. Their objectives were to review what is known of the seismicity in northern Canada, and to comment on the spatial relationships between seismicity and other geological and geophysical features. The data consisted of events monitored since 1962, the first year a large enough number of seismographic stations were maintained in the area to provide reasonably comprehensive coverage. They state that the better-defined epicentres are probably located to within 50 km, with the worst having a location error of up to 100 km, depending on the distance to the nearest stations. They then attempted to relate the clustering of these observations to various other features and concluded that they may be explained by known deformational features, by sedimentary loading of the crust and by residual disequilibrium due to crustal unloading at the time of the last major glacial retreat.

In evaluating the impact of the computer on the types of studies mentioned here, it is clear that for the magnetic and gravity surveys the computer is a necessity to handle the many thousands of data points acquired. Indeed, it is probably true to say that the ability to acquire large quantities of data grew from the same technology as, and because of the availability of, the ability to display the resulting information. Thus without computer-based mapping the projects in their present form would not have been performed. In the seismic reflection and earthquake incidence studies sophisticated instrumentation was needed for data acquisition, but the projects could have been completed without the use of computer cartography.

GEOCHEMICAL MAPS

Geochemistry, particularly in mineral exploration, is one of the oldest geological skills. Agricola, for example, in 1546, described six basic flavours of spring water, that were of use in determining mineral deposits. Indicator plants—species that grow near particular types of mineralization—have been known for a millennium. Placer exploration—panning for gold, tin, etc.—is also ancient. Geochemical exploration today is little different in this respect—elements are searched for in rock, soil, sediment, plant, and water samples. The main differences are in analytic instrumentation, sampling methodology and data processing of the results. It is not commonly realized how many geochemical analyses are, in fact, performed with modern rapid analytic instrumentation. In the 1971–72 year over 800,000 samples were analysed for at least one element in Canada, and over 300,000 in the U.S.A. Soil, rock, and stream sediment samples comprised the vast bulk of these—rock samples being more popular in the U.S.A. and soil samples in Canada, presumably due to northern terrain types. Most of the other analyses were performed on water and vegetation samples.

Sample *collection* has several aspects. It is the job of the geochemist to determine the best position in the stream bed or the best depth in the soil profile to collect a sample. This should be based on his knowledge of the dispersion mechanisms of the element or mineral for which he is searching. Another aspect concerns the spacing of samples—clearly, if samples are too far apart the target anomaly may well be missed; and if samples are too closely spaced considerable money is being wasted. A decision must also be made as to what kind of sampling to perform—it may well be much cheaper to sample vegetation for elemental analysis than to dig a suitable hole for soil sampling. The number of elements to be *analysed* in each sample is another factor in the economic equation, since several elements may correlate well together, or certain element combinations may indicate either a false anomaly (where the accumulation of the target elements is merely due to the dispersion agents, such as ground water) or strongly suggest a true one.

The *manipulation* and *display* of data after sample acquisition and analysis is the final step. As might be expected from the cost of data acquisition, considerable effort is sometimes expended on this stage but, conversely, a surprisingly large number of surveys are satisfied with elementary processing techniques. This is, in part, explainable by the lack of consensus on methodology. In addition, survey objectives may be either regional geochemistry, where the objective is to observe the general behaviour of some component over appreciable distances, or exploration geochemistry, where the objective is to determine anomalously high concentrations of some mineral of economic interest. Since the concentration of the target material frequently varies greatly over even small distances, appreciable smoothing should be performed prior to map production. For regional analysis, trend surface methods (see Davis, 1973) are commonly used for computer-based estimation of smooth regional trends. In exploration geochemistry careful examination of the possible population distributions is necessary before areas possessing appreciable concentrations of the desired material can be satisfactorily distinguished from the 'background' range of expected data values (Govett et al., 1975).

The first requirement of data examination is to determine the background and threshold values for the region. These are defined as the average regional value and the upper limit of regional variation exclusive of anomalous zones. This raises two points. First, a region has to be defined within which the background variation is more or less homogeneous (this may be identifiable from other information, such as rock type). Second, an anomaly is defined as having values outside (usually above) the expected background variation, so the argument is circular—as indeed it is for defining regions or domains as well. This is a major problem in cartographic work in geology—the delineation of zones with similar properties. Determination of regions or anomalies is not usually a major problem when mapping by hand—other aspects of the data can be taken into account by the trained geologist. The advent of computer analysis, however, rapidly demands the introduction of computer-assisted mapping, with a consequent demand to specify the criteria more precisely. In theory this is good, since it forces objectivity. Nevertheless, in practice it has merely deterred the use of the computer on a production basis, since it appears to the untrained individual that much effort and statistical skill is needed merely to simulate the 'obvious' manual interpretation.

The above remarks apply to both data posting—where anomalous values may be flagged automatically—or for contour maps. In the second case production of a map honouring all data values is frequently undesirable, as the errors in sampling and analysis may generate surface fluctuations that obscure the subregional behaviour. In statistical terms, the correlation between neighbouring points is not very high. For most purposes, it is necessary to filter or smooth the data.

Data smoothing occurs every time the mathematical surface generated from the data fails to honour, or pass through, all the input points. This is a natural consequence of the gridding stage in most computer programmes, since the requirement that the surface passes through the interpolated values at the nodes of what is usually a relatively course intermediate grid precludes the possibility of their passing through the data points, especially if several occur in one cell. It is necessary in most cases to be able to control the extent of smoothing, and this may be achieved using various approaches. For regularly spaced data on a grid, Robinson and Charlesworth (1975) used a Fourier filter to eliminate both excessively high and low frequencies. Their information was concerned with subsurface stratigraphy in western Canada, and for oil exploration purposes they wished to examine features a few kilometres in length that could possibly be buried reefs. Unfortunately controlled smoothing of this type is not readily obtainable for irregularly spaced data.

Contouring is usually performed using an arbitrary grid over the map area and estimating elevations at each node. Clearly enlarging the grid spacing increases the smoothing of the data. This necessitates some interpolation method to estimate grid node values from nearby data points. A typical method is the moving average, where the value at any location is a weighted average of the nearby points, closer points being weighted more heavily than more distant points. Two sources of deviation of the surface from the data values should be noted. First, if the weighting function is such that at any data location any other data points contribute to the surface being generated, then the surface will not pass precisely through the original data. This depends on the weighting function itself. A weighting of $1/d^2$, where d is the distance between data location and grid node, will not produce any smoothing. A weighting of exp $(-d)$, on the other hand, will do so. If $d = 0$, the first case yields an infinite weighting, while the second gives unity. Secondly, the very existence of an intermediate grid smooths the data uncontrollably, since data points not lying at grid nodes will rarely be honoured.

To meet the need for smoothed maps two competing techniques have been used in geology: trend surfaces and kriging. Trend surface analysis consists of fitting a polynomial of fixed order in x and y to the data, using conventional regression techniques. This has the advantage that traditional statistical techniques may be used, although it is assumed in the method that local fluctuations (the 'error' component) are uncorrelated. This is reasonable for errors in measurement of a geochemical variable, but is not correct for real small-scale or local variations. In addition, there is no foolproof way to determine how high an order polynomial to use. Figure 8.4A shows a third-order trend surface map of nickel distribution in soil samples in Sierra Leone, and Figure 8.4B shows the same data contoured using moving average techniques.

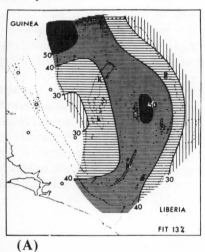 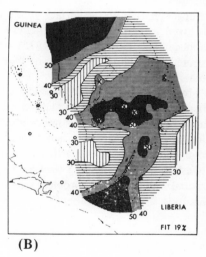

(A)   (B)

LEGEND

o OTHER ROCKS
  BASEMENT GRANITE
K KAMBUI SCHIST
  < 10 ⎫
  < 10 ⎭ CONTRAST
● CONTRAST > 5.0
O CONTRAST < 0.2

0   20   40 MILES

8.4 Regional distribution of nickel content in soil samples over the basement complex, Sierra Leone: (A) cubic trend surface; (B) moving average surface (Nichol *et. al.*, 1969; reproduced by permission of Economic Geology Publishing Co.)

The usual alternative to trend surface mapping is universal kriging, which is a form of moving average—that is, the value at a grid node is a weighted average of some neighbouring points. Several preliminary stages are necessary in order to generate these weights. The first step is intended to handle precisely the point that trend surfaces cannot—that points close to each other in space are likely to correlate strongly with each other. For a given set of data (and, by implication, a single domain) the variation of this correlation is examined with increasing distance between samples. Clearly this correlation is large between close samples, and almost zero for widely separated ones. This effect—expressed on a diagram called a semivariogram—is the primary step in kriging. Obtaining a good semivariogram requires appreciable experience, since it is necessary to fit a suitable mathematical expression to the correlations observed in the data. If this correlation is unity at zero separation distance then no error component is associated with data measurement. If the value is less than unity then some error in measurement occurs that is, two samples at the same location would not be entirely correlated with each other—the 'nugget effect') and the surface would be smoothed. An introduction to map kriging is given in Davis (1973). Journel (1975) describes some approaches used in other applications. Govett et al. (1975) provide an excellent description of sampling strategy, relevant frequency distributions and anomaly detection for geochemical data.

In conclusion it could be said that the advent of computers has been a mixed blessing. While technological development has dramatically increased the number of analyses performed, it has not produced a satisfactory 'hands-off' display technique for the field geologist who is not comfortable with statistical principles in general and hotly debated ones in particular. Underlying all the statistical approaches is the question of distinguishing between local and regional variation. This is in essence a spatial (and perceptive) problem—a point that must not be forgotten in any statistical analysis.

FACIES MAPS

The object of many geological studies is to classify an area into several facies (zones, domains) each with relatively homogeneous properties. The initial information consists of a set of point samples analysed for various constituents. A facies is defined as a laterally continuous unit, a unit being expressed as any identifiably similar combination of properties. Thus we have two problems: how to classify a set of observations into several facies or categories, and how to display the end product. In many cases, facies are somewhat arbitrary subdivisions of a continuum, although occasionally sharp boundaries between units may exist. An overview of a wide variety of manually produced facies maps may be found in Forgotson (1960).

One may wonder why continuous variables should be broken into distinct categories, rather than being contoured as a continuous surface. The answer is that we may normally contour only one variable per map, and a sea-floor environment, for example, consists of many interrelated variables. Statistical classification schemes such as cluster analysis may be used to help define several distinct groups of samples such that the within-group similarity is greater than between-group similarity. An alternative scheme is to use factor or principal components analysis in an attempt to generate one or more composite variables (consisting of varying proportions of the analysed variables), that express most of the overall variation. Each composite variable or factor may then be considered as continuous over the map area and contoured by conventional techniques.

These numerical techniques are all relatively recent, whereas facies maps have been constructed for many years. More subjective categories were used before the widespread availability of computers permitted extensive mathematical computations. Indeed, much of geology consists of this classification of continuously varying material into artificial but meaningful categories. It is still useful to describe a rock as a granite even if there are cases where the differences between it and other rock types are artificial or even vague. Thus some traditional 'geology maps' (or, more correctly, lithology maps) may have boundaries drawn between continuously gradational rock types. Often, however, the boundaries between the geographical occurrence of various rock types may be defined fairly precisely. The techniques described below are fairly widely used computer methods that have replaced, in some part, the intuitive drawing of facies boundaries from the tabulated or plotted analyses.

In the numerical processing of facies information the data are initially edited for errors. They are then usually standardized so that each variable has a mean of zero and a standard deviation of one. This step is used to prevent one variable from dominating subsequent operations merely because of large measured values.

The next stage, data reduction, is performed to reduce the number of variables used in further processing. In many situations some of the parameters that are measured may well correlate strongly with each other, as they are merely expressions of a few underlying forces. As an example, in a shore environment the aspect of the shore to the prevailing wind and waves may affect particle grain size as well as faunal species and abundance. Thus many of the measured variables may be redundant in that, individually, they add little to the description of the total variability of the environment. The main methods used for grouping variables are R-mode principal components analysis and factor analysis. Both of these methods, which are very similar in application, are used to generate a set of components or factors that are linear combinations of the initial variables. These are frequently obtained so that the first of these composite variables explains as much of the total variation as possible, the second as such of the residual variation as it can, etc. Consequently only a 'few' are required to explain 'most' of the original variation in the measured variables. Deciding how few, however, is not always easy.

The input to either of these processes is a matrix expressing the similarity of every variable to every other variable. This is often the classical product–moment correlation coefficient, but it need not be. The difference between principal component and factor analysis lies in the underlying model. In the principal components approach it is assumed that the components being extracted can be described entirely by the measured variables with no need to consider other properties that could have been measured but were not. In true factor analysis, allowance is made for the incomplete description of the factors by the actual variables measured. This depends on a variety of assumptions about the data, and the various methods of attempting this feat will not be described further here.

In many cases the number of properties describing the data variation can be drastically reduced by the techniques described above, and it is frequently advantageous (although not obligatory) to perform an operation of this type to aid in comprehension of the data.

Most techniques for grouping samples (Q-mode analysis), rather than variables, require as input a matrix containing similarity measurements between these samples. Various possibilities exist for this similarity measure—those like the product–moment correlation coefficient that increase (from -1 to $+1$) with increasing similarity, and those like Euclidean distance that decrease to zero with increasing similarity. Preliminary reduction in the number of variables can provide an appreciable saving at this step, as most grouping methods increase in cost approximately with the square of the number of samples.

Given a similarity matrix, there are commonly three approaches used to group the data. The first is Q-mode principal-components or factor analysis. This is simlar to the R-mode methods just discussed, except that the factors or components extracted are expected to be associated with clusters of similar *samples*, rather than *variables*. Alternatively, depending on the methodology used for rotating the factors, they may be considered to be theoretical end-members of the suite of samples. While Q-mode component or factor analysis is fairly common, it tends to increase in cost very rapidly with the number of samples.

A second technique is discriminant analysis, although this method is intended for somewhat different initial conditions. A set of samples is defined as falling into two or more groups and a set of 'discriminant functions' is obtained to distinguish as well as possible between them. Subsequent samples may then be classified into groups using these derived functions. In many mapping applications, however, initial categories and classified samples are not available, necessitating the generation of sample groups from the measured variables alone.

The third common technique is cluster analysis. This term encompasses a wide range of usually non-statistical grouping techniques. Given the similarity matrix described above the next step is to arrange the samples, usually into a hierarchy, so that samples with the highest similarity are placed together. These groups are then associated with the groups that they most closely resemble and so on, until all of the objects have been placed in the classification.

Just as there are many ways of generating a similarity matrix, so also there are many ways of clustering the groups. Most of the common methods of both are described in Wishart (1975).

Jaquet *et al.* (1975) collected seventy-six samples from the sediments on the floor of Lake Geneva (Figure 8.5A) and analysed them for twenty-nine chemical components. The data were standardized and a product–moment correlation matrix was obtained for each variable against each other variable. Data reduction was

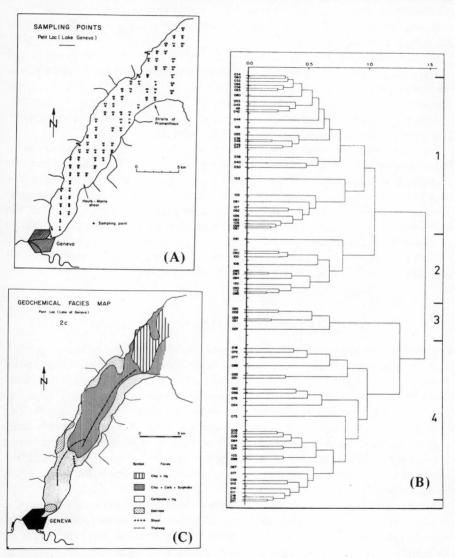

8.5 Petit Lac (Lake Geneva) sediment samples: (A) sampling locations; (B) dendograph showing sample clusters; (C) geochemical facies map (Jaquet *et al.*, 1975; reproduced by permission of Plenum Publishing Corp.)

performed using both principal-components and factor analysis techniques to produce four composite variables. These four variables were then used to generate a sample-similarity matrix using Euclidean distance as the measure. Clustering was performed using the 'unweighted pair-group' method, and displayed using the dendrograph approach of McCammon (1968). An example of this is given in Figure 8.5B, which was derived from four standardized principal components. The resulting facies map is given in Figure 8.5C.

Having evaluated various combinations of procedures they suggest that standardizing variables, followed by principal-components analysis for reduction in the number of variables, gave the best results with the fewest statistical assumptions. The effects of varying the clustering approach were not examined. In the end, however, it is the interpretability of the resulting map that is of greatest importance, and most of the techniques used produced plausible maps.

Facies mapping by computer suffers from advantages and drawbacks similar to those of computer-assisted geochemical mapping, with the additional problem that costs increase dramatically with large numbers of samples. Again, spatial closeness is rarely taken into account in the grouping of samples. For a small number of samples with many measured variables automated sample grouping is commonly used, but this is done less frequently when a few parameters are measured on many samples. It is primarily a matter of convenience whether computer mapping techniques are used for data display—partly because little work has been done on techniques for generating boundaries between groups of similar samples.

ORIENTATION DATA AND DENSITY MAPS

Orientation data, a fairly common type of geological information, includes properties such as water runoff direction, flow direction based on current ripple marks, direction of ancient glaciation based on existing abrasion marks, pebble orientations, and rock fractures. The orientation may involve a compass direction only, or it may involve a dip component. It may possess merely an orientation, as with pebble long axes, or it may also possess a direction, as with water runoff. Appreciable difficulties may be encountered in manipulating this kind of data—calculating a conventional average is not trivial when 360 degrees is the same as zero! Much of the research into this kind of data has been performed by geologists (for example Ramsden, 1975; Ramsden and Cruden, in press).

Problems arise in the field work or collection of orientation data because it is typically highly variable, and several local observations of pebble orientation, etc., must be made in order to derive a reasonable average direction. The local results are frequently displayed on the map as rose diagrams, these being circular histograms displaying the number of occurrences of an orientation within each of several sectors. Nevertheless, appreciable difficulties arise if it is desired to display flow directions, for example, as a continuous property over the map area. The mathematical difficulties are described in Agterbeg (1974), and the cartographic problems are not yet fully resolved.

When the orientation information can be considered to be internal to a body of rock it is referred to as a fabric; fabric being defined as the internal geometric properties of the body. These properties are the cumulation of individual structures having geometric orientation—for example planar joints; bedding or cleavage; and linear features such as the long axes of sedimentary particles or the orientation of ripple marks. Each structure type with a distinct geometric configuration is called a fabric element, and the overall fabric is the sum of all these fabric elements.

One object of fabric analysis is to describe the fabric of the geological body under study, and a major aspect of any such study is the need to subdivide the rock body into a series of spatially distinct portions or domains within which the fabric may be regarded as homogeneous. The selection of fabric domains depends on the size of the smallest portion of the body that may be considered a distinct unit—that is, it depends on the scale of the study. The normal procedure for fabric analysis is to collect data as uniformly as possible over the study area; divide the area into domains visually; plot orientation diagrams for each domain; and then combine any domains that appear similar. It is also possible to contour angular dip measurements, and regions of homogeneous fabric are identified as regions of low contour density.

Ramsden (1975) described a computer-based procedure in which the area is divided into sampling units by the geologist, and the data are treated throughout as three-dimensional, eliminating the need to examine horizontal and vertical components separately. Statistical tests are used to indicate domains with large scatter of orientations, and fabric diagrams are automatically produced. These items are primarily aids in separating the study area into meaningful domains, within which the fabric is basically similar, and between which the fabrics vary. Maps may be produced which indicate the orientation of individual readings by a line whose own orientation indicates the orientation of the sample, and whose length indicates the vertical component (Figure 8.6A). Once domains are defined, the vector mean may be displayed for each (Figure 8.6B). Alternatively, the deviation of each domain mean from the regional mean may be displayed (Figure 8.6C).

In the process of identifying domains it is frequently desirable to display the scatter of all the samples as an aid in interpreting fabrics and in locating relatively homogeneous regions. This is usually handled by considering all the samples within some region to be located at a point, and the measured directions or vectors to be projected onto a hemisphere or sphere. The next step is to contour this surface in units of point density and examine the result for various clusterings of orientations, using statistical techniques if possible. The standard approach is to define some counting area and move this over the map, determining how many observations fall within it at each location. Unfortunately, the resulting map varies drastically with changes in the counting area—from a very smooth map if the areas are large to a delta function (zero where there is no sample, unity where there is) as the counting area gets infinitely small.

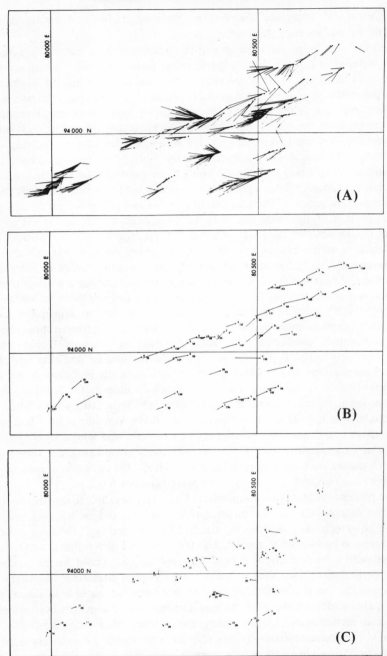

8.6 Normals to observed bedding orientation: (A) raw data; (B) grouped data; (C) residuals from overall mean (Ramsden, 1975; reproduced by permission of J. Ramsden)

Ramsden (1975) shows that the common methods of density estimation can be generalized to the form

$$d = (1/n) \sum_{i=1}^{n} w(\theta i)$$

where n is the number of samples, i is the distance of the sample point from the counting location, and w is a weighting function. For the constant-area method (where a point is counted if it falls within a circle of radius c), $w(\theta i) = 1/a$ (the area of the counting circle) if $\theta i \leq c$, or zero otherwise.

An alternative weighting scheme is to apply, instead of a constant weight $1/a$, a weight decreasing regularly with increasing θi. If the Fisher distribution is used, $w(\theta) = (k/2 \sinh(k)) \exp(k \cos \theta)$, where k is a suitably-chosen concentration parameter. A third method is to adjust the area of the counting circle until a coefficient of variation reaches a prescribed level, and the constant-area equations are then used.

Ramsden examined these methods in detail with respect to the selection of suitable parameters. He concludes that there is no basis for selecting any set of parameters (that affect the degree of smoothing of the surface) if the model (that is, the expected number of clusters, or density peaks) is not known. Given an expected number of peaks, however, some selection of suitable parameters may be made. This is in accord with work done on conventional contour mapping, where it has not been found possible (on the basis of the data alone) to distinguish between data points that are poorly correlated because they only occur at each peak and pit in the topography (that is, at or close to the Nyquist frequency, or sampling limit); poorly correlated data due to the (accurate) measurement of a surface having features of frequencies higher than the Nyquist frequency; or data having large errors in the x, y, or z measurements.

While much of the work on orientation data has been inspired by the availability of the digital computer, there is no overwhelming need to use the computer for map production. Nevertheless, as a matter of convenience most data display will probably be automated, since engineering geology problems can generate reasonable quantities of data, and domain definition tends to be an iterative (and hence repetitive) process.

GEOLOGICAL AND LITHOLOGICAL MAPS

The classical 'geological' map displays the rock types present in a region, the level of tectonic activity, or some similar attribute. It is concerned with variation in three dimensions, but with sampling limited to two in most cases. They may be of two types: area coverage maps showing which rock types are to be found in any particular area; and 'posting' maps indicating the values of particular parameters at observed locations. These values may be the rock type observed at each outcrop; the orientation of any faulting; or any of a wide variety of possible parameters. The

8.7 Geological field data sheet (Berner et. al., 1975; reproduced by permission of Geological Survey of Canada from Paper 74-63)

normal procedure is for field forms (such as that of Figure 8.7) to be processed, and postings made of interesting parameters. On this basis the geologist will construct his area coverage map. It should be noted that his boundaries usually indicate more than merely some mean distance between samples of different types—the *form* of the boundaries frequently indicates the structure, such as faulting or folding interpreted by the geologist to explain the observed information. For this reason it is, and will probably remain, difficult to produce good final geology maps without considerable manual intervention.

Geological rock types may well be sampled in three dimensions if it is of economic interest to do so. Oil company well logs are a prime example. In this case the three-dimensional contacts must be defined, then displayed in two dimensions. Contour maps and cross-sections are frequently used; the one to display the overall behaviour of a single contact, the other to display the behaviour in a limited direction of all of the geological surfaces of interest. The two-dimensional limitations of a map sheet are a great inconvenience in this type of work, and three-dimensional wood or foam models are sometimes used. Many challenging problems remain in displaying this type of data.

Even within the realm of the most traditional of map types—the lithology map—much development work is possible. Two recent examples are the work by Bouillé (1976, 1977) and Burns (1975). Both are concerned with the systematic description of the relationships between delineated zones on a map sheet. These relationships may then be manipulated by computer.

The work of Burns concerns the fact that geological events occur in a time sequence, and any historical understanding of the geological processes requires the derivation of this sequence. Unfortunately it is usually only possible to determine the age relations of rock types where pairs of them meet. Figure 8.8A shows a schematic geological map of an imaginary area, possessing lithology types A to L. Arrows indicate contacts between lithology pairs at which relative age determinations may be made on the basis of superimposition, intersection, etc. The arrows point from older to younger rocks. Figure 8.8B indicates the processing of this information. The upper left diagram records the younger member of each lithology pair in the relevant cell. The upper right diagram shows the same matrix reorganized by switching rows and columns until older events are to the left of or above younger events. The lower left diagram shows the result when all other relationships deducible from the previous set are filled in (that is, all columns above any particular entry are filled with the entry value). Finally, the lower right diagram indicates the deduced sequence of events, starting with LDFH and finishing with JCBG. Due to the incompleteness of the final matrix the relations of K and AE to each other are unknown. Reference back to Figure 8A indicates a discontinuity or fault in the top centre. This clearly occurred after deposition of units K, H, F, and D which are dislocated, and before the occurrence of J and A, which are not. This information is sufficient to date K as prior to A, and hence the correct sequence of events has been deduced from the geological map.

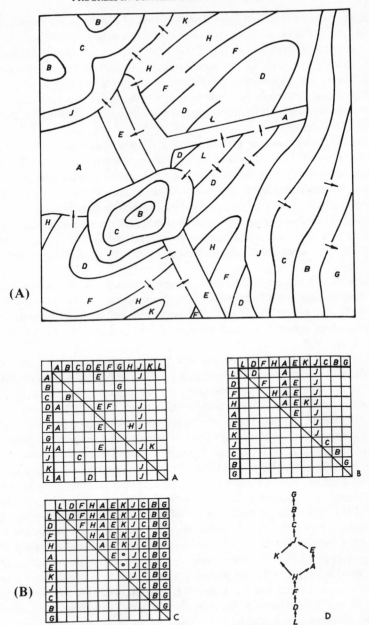

8.8 Sequence of geological events: (A) rock types with older/younger relationships across boundaries; (B) derivation of event sequence (Burns, 1975; reproduced by permission of Plenum Publishing Corp.)

This procedure of Burns is a somewhat more elaborate version of what, in computing science, is called a topological sort, for which a simple algorithm is given in Knuth (1968). This approach assumes that any legitimate sequence is satisfactory, as is the situation of the top right of Figure 8B, where sequence K and AE could legitimately be interchanged. The concept of a topological sort is an extremely valuable one in many spheres of computer-assisted cartography. A good example is whenever a series of objects or polygons need to be ordered 'front to back' for perspective block diagrams or other applications, and only the relative positions between adjacent objects are known. A development of possible interest is the work of Gold and Maydell (1978) in which a region is defined as a set of triangles, possibly with nodes representing the locations of objects, or else with polygons defined by one or more triangles. A simple algorithm is given to achieve a front-to-back ordering from any direction for any triangulation, merely by starting at the foremost triangle and processing neighbouring triangles in turn. The method also works for a radial expansion out from any specified point.

The work by Bouillé on geological maps is also concerned with adjacency relationships, but is based heavily on the availability of SIMULA 67, a very high-level programming language which permits extremely flexible data structures. He has developed an extension to this, HBDS (Hypergraph Based Data Structure), to express the graph-theoretic relationships between geological entities that are frequently displayed as geological maps.

One example of this involves a very similar problem to that of Burns in that a stratigraphic summary is derived from the map, but in this instance the initial problem was to digitize a lithology map so as to facilitate subsequent computer operations. First of all, the nodes (junctions of three or more lines) are digitized and numbered, and then the arcs are digitized between nodes. Associated with the arcs must be the two node numbers and the left and right stratigraphic units. From this a graph (in the mathematical sense) is constructed. A dual of the graph is also constructed. If the graph is thought of as a set of countries with irregular (digitized) borders, the dual of a graph may be visualized as a road network connecting the capitals of each country (located anywhere within its borders) to the capitals of each adjacent country.

This dual graph expresses all the neighbourhood relationships between the different stratigraphic units (or, more correctly, the many possible separate areas of each stratigraphic unit or rock type). From this a stratigraphic summary graph (Figure 8.9A) may be derived, retaining one linkage between each pair of stratigraphic units which were linked on the dual graph. From the summary graph various properties may be observed:

> an arc between a vertex i and a vertex $i-1$ shows a normal stratigraphic situation;
> an arc between i and j ($j \neq i+1$ or $i-1$) implies an unconformity;
> on the contrary, a lack of an arc between i and $i-1$ does not imply an

unconformity, but shows that connection between i and $i-1$ is nowhere observable on the map; and

lack of arc between a vertex i and any other tops indicates that in the present situation, the layer 'i' is known only from drilling, (Bouillé 1976, p. 380).

Other work by Bouillé is more generalized, his HBDS (Hypergraph Based Data Structure) language being based on set theory and the hypergraph.

It deals with four basic abstract data types which are named CLASS, OBJECT, ATTRIBUTE and LINK. They respectively represent: set, element, property, and relation. But, these fundamental concepts of the set theory are always available and we then may define and manipulate sets of classes, sets of relations, relations between sets of classes, etc. (Bouillé, 1978, p. 2).

This system has been used for a wide variety of problems, primarily of a cartographic nature, since the early work on geological maps described above. Examples include hydrography and hydrographic networks, stream systems, road networks, administrative divisions, and urban development. Two examples only will be given briefly here. Figure 8.9B shows a possible data structure for a simple topographic surface. Various classes have been defined: map, contour level, curve, summit, and reference point. The diagram shows the relations between elements as curved lines. Of special interest are the heavy lines. These form a skeleton to the whole data structure, even in the absence of any data, as they show the relations between classes. With a data structure of this type a wide variety of otherwise difficult questions may clearly be asked.

A final example of the flexibility of the concept is the draughting of a perspective drawing of a map composed of contour lines and a road network:

The method consists of: starting from the highest level and considering the set of its corresponding isolines, drawing them, drawing the parts of the segments which are included in these curves; then we consider a lower level, which is drawn taking the windowing into account; then the parts of the segments intersecting these curves or included between them and their possible upper curves, are also drawn in the same manner; we thus do likewise with the lower level until the last one ... (Bouillé, 1978, p. 21).

Perhaps the most amazing aspect of this example is that a link between two classes need not be a pointer, but may be an algorithm—an intersection test in this case! With this approach the handling of cartographic problems may not always be efficient, but is almost limitless. While the handling of lithology maps by computer is

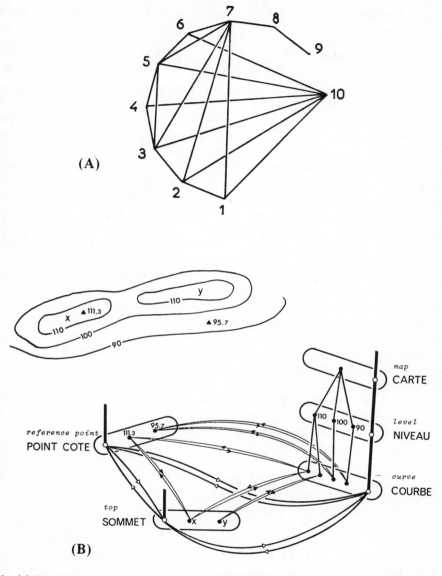

8.9 (A) Stratigraphic summary graph of relationships between layer types 1–10 (Bouillé, 1976; reproduced by permission of Plenum Publishing Corp.); (B) data structure representing some topographic features (Bouillé, 1977; reproduced by permission of President and Fellows of Harvard University)

still in its infancy, valuable work has already been done in defining some of the ground rules.

TOPOGRAPHIC MAPS

Contour maps are widely used to represent *surficial* topography. Geologists often need to proceed beyond this and map *buried* topographic surfaces. These may well have once represented the earth–air contact (or, more commonly, the earth–water contact) but they have been buried and frequently deformed by folding or faulting. Very large sums of money are spent each year by oil companies in collecting and processing this type of data. Consequently computer contouring techniques have been developed by many individuals, and many maps of this type have been and are being produced with varying degrees of success. It should be noted that certain differences exist between subsurface and surface geological mapping because of the differences in the properties of the data used. The properties of a data set for topographic modelling can be described (Gold, 1977) under the heading of sample content, sampling adequacy and sample isotropy. Under sample *content* would be considered the sample's x–y location information as well as any elevation or slope information associated with that location, and error estimates associated with each of these. Photogrammetric methods will not generally produce slope information, whereas some borehole techniques will do so. The x, y, and elevation values of a borehole data point will usually also be relatively precise.

The biggest differences between these two data types are noticed when considering sampling *adequacy*. Photogrammetric methods generally produce from tens to hundreds of thousands of data points per stereo model, whereas, due to the costs, a few tens of sample points must suffice to define a buried surface delineated by drilling. Sample point density must be adequate to resolve features that are of sufficient size to be considered important. Current topographic surfaces often pose no problem, but frequently in subsurface work the sample frequency is less than desired.

The third data property, sample *isotropy*, has rarely been considered, but is an important technical consideration in many practical mapping problems. It concerns data point distribution over the map sheet. Davis (1973) classifies data point distributions into regular, random, and clustered. He describes various tests used to examine data point distributions, primarily using techniques either of sub-area analysis or nearest-neighbour analysis. The results of some techniques, for example trend surface analysis, are heavily dependent on the uniformity of the data distribution.

Data point isotropy, however, is more concerned with the variation of data point density in differing directions. For most randomly collected data sets the anisotropy is not marked and can be ignored. What is often forgotten by theoreticians is that a very large quantity of data, particularly that acquired by automated means, is collected in strings or traverses, where the sample spacing along the traverse is much smaller than the distance between traverses. Many very important data types fall into this category, including seismic profiling, ships' soundings, and airborne surveys of many types (magnetic, gravimetric, radiometric, etc). An interesting data type at the research level consists of the input of contour lines digitized from topographic or other maps. A topographic map may only be considered a general topographic model if its data structure (contour strings) may be converted to some other form (for example, regular grids of elevations) for further study. In addition, many non-automated geological data collection techniques consist of field traverses—along roads, streams, ridges, or valleys.

Three aspects of this discussion are worthy of further note: the concepts of a topographic model; of data collection along relevant features; and any special processing requirements of traverse data.

The Topographic Model

In computer terms at least, a topographic model should be distinguished from a topographic map. As with a balsa-wood or styrofoam model, it should be viewable in many ways—from any orientation, by slicing it, etc. A topographic contour map is merely one way of displaying the topographic model. The primary requirement for any display of the model—contour map, cross-section, block diagram, etc.—is that the model may be interrogated to obtain the elevation at any x–y location, and that this value should be obtained in some reasonably efficient manner. Since there will not usually be a data point precisely at each desired location, a topographic model should be defined as a set of data points *plus* the required algorithms to obtain any requested elevation. Most modelling or contouring algorithms assume that the data point location has no intrinsic meaning. *This is often not the case.* Where the surface is visible in advance (for example, current topography, but definitely not buried topography), some data points at least would be selected to occur at peaks, pits, ridges, or valleys. In many cases, samples would be selected along a 'traverse' of a ridge or valley. Peucker (1972) calls these items 'surface-specific lines or points'. While manual mapping methods may take breaks in slope into account at these locations, this is difficult with automated techniques.

Because of the widely discrepant distances between data points for traverses (which need not be perpendicular to the x or y axes) some modelling methods may break down. This is typically true of interpolation techniques that perform some form of weighted average on the 'nearest' few data points they can find in order to evaluate the elevation at some unknown point. Obviously, having to search the whole data set to find the nearest few points to each location on a grid, for example, may be time-consuming. In addition, the nearest few points will often be obtained from one traverse exclusively, giving no weighting to values from adjacent traverses that would provide information on the behaviour in the second dimension. Many programs acquire 'neighbouring' data points until values are obtained in at least six of eight sectors of a circle around the unknown point. This is an attempt to compensate for the anisotropic distribution of sample locations around the point of interest, but for traverse data this requires much computational effort and frequently little improvement is observed.

The problem arises because a simple metric distance is a poor measure of neighbourliness, especially in anisotropic cases. A non-metric definition is needed. This may be achieved by triangulation techniques (Gold, 1977; Gold et al., 1977; Males, 1977) whereby some 'best' set of triangles is defined to cover the map area, with all vertices at data points. Neighbours may thus be found in a fashion independent of the relative data point spacings and without the necessity of performing a search for neighbours at every grid point.

Topographic mapping is used by many disciplines; what is different in the case of geology? Our use of conventional topographic survey maps is much the same as anyone's, except perhaps for a greater need to extract specific elevation values from the map when determining sample locations. Isopach, or thickness, maps are common; these may be considered as topographic maps of the top of the geological unit, the base of the unit being taken as zero.

A wide variety of contouring programmes have doubtless been used by geologists to construct isopach maps. Most of these programmes require the generation, from the irregularly spaced data, of a regular grid of estimated values. The most common procedure would be to extract the thickness information directly from each sampling or drill hole location by subtracting the elevation of the bottom contact of the geological unit from the elevation of the top contact. The thickness values at each data point are then contoured by estimating the values at each node on a grid, and interpolating contour lines within each grid square.

While this approach produces a reasonable map where the geological unit is continuous, shortcomings become apparent when the stratum is absent in places. In particular, the zero thickness contour line is frequently implausible. This is the result of contouring a computer-generated grid with zero values in regions where the unit is absent, and positive values elsewhere. The interpolated zero contour will therefore follow the grid square edges.

It is clear that the thickness 'model' is inadequate and that zero thickness is not necessarily an adequate description of the absence of a geological unit in a particular location, specially if the absence is due to the erosion of pre-existing material. It is therefore more correct to consider an isopach map as a map of the *difference* in elevation *between two complete topographic models*, one of the upper contact and the other of the lower contact of the geological unit. With this approach, negative values are legitimate and problems with the zero isopach line disappear.

An interesting example of this type of mapping concerns a coal resource evaluation project in the Foothills of the Canadian Rockies (Gold and Kilby, 1978).

Coal resource evaluation clearly requires information on the mineable tonnages, as well as the grade, of the coal strata. In open-pit mining the volume of coal economically extracted from a seam is directly related to the amount of overburden that must be removed in order to expose the coal. This may very conveniently be expressed as an 'overburden ratio' map, being a contour map of the ratio of the overburden thickness divided by the coal seam thickness at the same location. While the cutoff value for economic recovery varies due to many factors, it is relatively uncommon for coal with a ratio of greater than 10 to be economically mineable. The overburden ratio map may thus be of value in coal reserve estimation.

The construction of an overburden ratio map requires three components: a topographic model, a model of the top of the coal seam, and a model of its base. In this study the coal seam was fairly consistent in its thickness (of about 30 ft) and therefore no model of the base of the coal seam was used for thickness estimation.

Once the topographic surface has been modelled, the same must be done for the top contact of the coal seam. This poses particular difficulties in that most of the data is obtained from outcrops and drill holes along one line—the intersection of the coal seam with the topographic surface. It is therefore necessary to estimate the geological structure (that is, folding) of the originally flat-lying sedimentary rocks, so as to permit projection of the seam perpendicular to the 'trace' of the coal. As in the topographic model, suitable mapping software and computer resources must be available. In addition, geological expertise is necessary to evaluate the geological structures, on the basis of the work of Charlesworth *et al.* (1976) and Langenberg *et al.* (1977), who discussed the criteria necessary in order to assume the presence of cylindrical folding within a domain—that is, under certain mathematical conditions, a particular portion of the coal seam may be satisfactorily described by a type of folding similar to a sheet of corrugated iron, linear in one direction and undulating perpendicular to this. If the orientation of the desired geological contact has been observed at several locations in the field, this linear direction, called the fold axis, may be determined by the mathematical techniques described in the previously mentioned references. A cross-section may then be constructed normal to this fold axis and all data points projected on to this 'profile'. The irregular folding of the coal seam in a portion of the study area is shown in the profile of Figure 8.10A.

In the absence of satisfactory orientation data it is possible to estimate a suitable fold axis in a trial-and-error fashion by examining the profiles generated for each estimate.

If a suitable profile can be generated for each domain, it is a suitable description of the geological contact. However, it is described at a different orientation in space from the original map, being perpendicular to the estimated fold axis. Dummy data points may then be generated on the map and rotated into the coordinate system of the profile. Their elevations may then be estimated from the profile, and the resulting points transformed back into the map coordinate system. These values may then be contoured by the methods previously described for the topographic surface (Figure 8.10B).

The final step in the procedure consists of generating the overburden ratio map from the *two distinct surface models*. The procedure is straightforward: the elevations of the two surfaces must be evaluated at a series of map locations (perhaps on a grid): the elevation of the top of the coal seam subtracted from the topographic

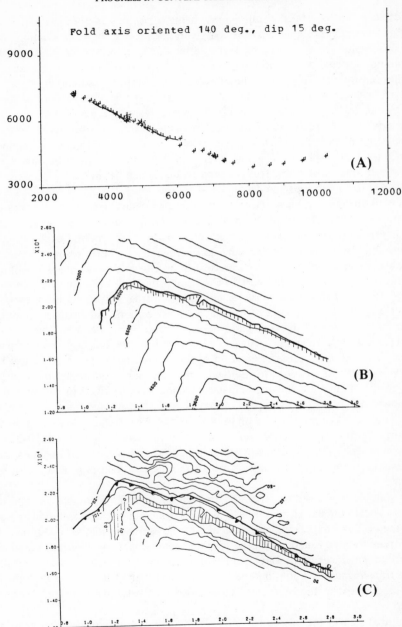

8.10 (A) Profile looking down fold axis; (B) Intersection of geological contact with topographic surface (existing contact indicated by shading); (C) Overburden ratio map (heavy line indicates faulting; shading indicates mineable coal seam (Gold and Kilby, 1978))

surface; the resulting difference divided by the seam thickness; and a contour map produced of the result (Figure 8.10c). These results may legitimately be negative where the coal has been eroded out, although normally only positive values would be contoured. On the basis of the local cutoff point for the overburden ratio, tonnages may readily be determined, either manually from the map or by further computer processing.

Two important conclusions may be derived from this example, and in the opinion of this writer they hold true for all map types—whether topographic or thematic. The first is that a *model* consists of an attempt to generate a space-covering map of a single parameter only. It should conform as closely as possible to the original data and avoid synthetic devices such as grids. The second point is that grids or rasters should be display devices *only*, to be used as needed to compare and display models. Ideally they should not be preserved or used again to make subsequent comparisons. The large effort expended to create good polygon-overlay procedures and to handle the resulting multitude of small residual polygons may, in some applications, have been better spent on generating good techniques of simultaneously scanning several separate polygon overlays and generating a final raster-type display.

The use of similar techniques for topographic modelling as well as for lithology maps introduces the final example in this section. It concerns a topic very close to the heart of many petroleum geologists—how well can a computer, as compared with a seasoned geologist, estimate the structure of a surface (a buried geological contact)? Dahlberg (1975) constructed an imaginary coral reef with several pinnacles, took random samples of from 20 to 100 elevations from the model and submitted them to two geologists and one computer (using a triangulation-based contouring algorithm). He showed that the computer-based model was always more precise, especially so with the fewest control points; so much for professional judgement! However, when additional geological information was available—such as general regional trends or depositional environment—the geologist's interpretation of many data sets improved dramatically.

COMPUTER-BASED DATA HANDLING

Data Capture

The following list gives the collection technique and estimated cost (in U.S.$) of each data point collected from a variety of surveys of interest to geologists:

Airborne magnetic ($2), ground gravity ($15–$20), marine seismic ($35), land seismic ($250), magneto-telluric ($800–$1000), shallow well ($50,000–$100,000), medium well ($250,000–$500,000), deep well ($1,000,000–$5,000,000). Clearly the quantity of information acquired for each technique varies dramatically. Nevertheless there are still a lot of deep drill holes in existence, since they are clearly of great commercial value. It is perhaps more useful to speak of the *density* of observations obtained as a function of the *area* of the *smallest feature of interest*. In most petroleum geology work the relative density of seismic results is much greater than that for drill holes. Each drill hole, on the other hand, provides a large number of different items of information. Data storage and retrieval techniques will consequently differ.

The aspect of data collection that has received the greatest published attention is the automation of geological observation in the field. This has traditionally involved the devising of suitable coding forms. The great problem with this is that, to date, computer systems are not nearly as flexible as field geologists whose *training* is to note any remarkable feature contributing to the overall aim of the study. By restricting him to any standard set of features to be described or identified, one is running directly counter to his training and philosophy—always a bad thing in systems design—as well as drastically reducing his real value in terms of relevant observation. Conversely, a check list is a useful memory aid, especially for subordinate-level field staff, although, like fire, it is a bad master even if a useful servant. The discipline of true observation, the persistence to truly look for clues to the problem in hand for outcrop after outcrop under frequent conditions of discomfort, is a hard thing to acquire with the best of incentives. It is nearly impossible when the view is dominated by a multiple choice questionnaire, the completion of which requires method but little imagination and whose final line irresistibly suggests that all needed information has been acquired. It should also be noted that a field survey is always problem-specific, and since a geologist would view the same piece of terrain very differently depending on the objective, a general-purpose coding form would probably be horrendously elaborate or hopelessly inadequate.

Nevertheless, much work has successfully been done using this approach. The greatest success has probably been achieved where a project has continued for several years with the same, or continually improving, coding forms and objectives (typically merely extending the area of coverage)—and where heavy dependence is put on seasonal or partly trained field staff. Other suitable applications are where the possible field observations are in any case limited—for example in drill-hole work, whether in shallow urban engineering–geology applications or deep, complex oil exploration. In conjunction with the availability of more sophisticated data base management systems, input of free text for each observation station has become more common, but some basic problems still remain in the editing and identification of relevant key words and phrases. Hutchison and Roddick (1975) conclude that such an approach requires less time on outcrop, loses less information due to illegible handwriting, and improves the consistency of recording information. The main disadvantages are the time lost in editing input data and in file and data base management. Garrett (1974) gives a good description of field coding forms for geochemical surveys, as used by the Geological Survey of Canada. These forms vary depending on the type of sample (soil, water, etc.) being collected. Clearly the requirements of such a survey type are simpler than those for general field surveys, but even here the bulk of the coding form is left available for comments or parameters defined by the project leader.

Another excellent example of the effort required to develop a comprehensive field data system, even for specific objectives, is found in the LEDA system (David and Lebuis, 1976) for Quaternary geology. Six computer-based forms are defined, each convertible to one 80-column punch card (Figure 8.11A). One additional form is reserved for sketches and photographs. Form 1 describes the sampling location, neighbourhood, and other relevant material. Form 2 describes a single stratigraphic unit at that location, including thickness, structure, texture, etc. Form 3 describes the composition of any stratigraphic unit or other geological entity, and measured parameters that may include fossil types, rock types, mineral types, etc. Form 4 (with the same layout) describes any orientation information, and Form 5 is designed to accept comments. Form 6 is used to record detailed measurements of the thickness of subunits occurring within a composite stratigraphic unit. These forms may be used in various legitimate combinations to describe any geological section, using a carefully selected and open-ended series of codes for each position on the form. Figure 8.11B shows, without need for further description, how the predefined form types may be used to describe an extremely complex geological section.

Data Base Management

Computer-based storage and retrieval of geological information has a long history (the earliest paper is by Margaret Parker, in 1946, on the use of punched cards in tabulating drilling data). Bibliographies up to 1972 include 447 references (Hruska and Burk, 1971; Burk, 1973). Some of the major geological topics covered include geochemistry (58), geological field mapping (47), geophysics (24), hydrogeology (19), hydrology (15), geophysical logs (19), mineral and fuel deposits (62), mineralogy (13), oceanography (13), palaeontology (47), petrology and lithology (46), stratigraphy (27), and petroleum and gas wells (61). The computer topics discussed include bibliographic files (91), bibliographies (10), data codes and coding (121), computer systems (110), curating (21), data display (117), data files (286), data recording (160), digitizing (15), indexes and indexing (69), information centres (36), national and international systems (53), data standards (48), surveys and questionnaires (14), theoretical topics (95), and thesauri (36). While much other work has been done, clearly a major effort has been made in the handling of computer files of geological data acquired in the field and intended for some form of geographical display.

The phrases 'data base' and 'data base management' are applicable to the handling, however primitive, of any collection of data, however small. However, in the computer business they imply the relatively sophisticated manipulation of relatively large data sets. Nevertheless, the types of manipulation can be described under the headings of collection, storage, updating, rearrangement, retrieval, display, and analysis. The first of these has already been described; the next three items concern the construction of the data base, and the last three concern its examination.

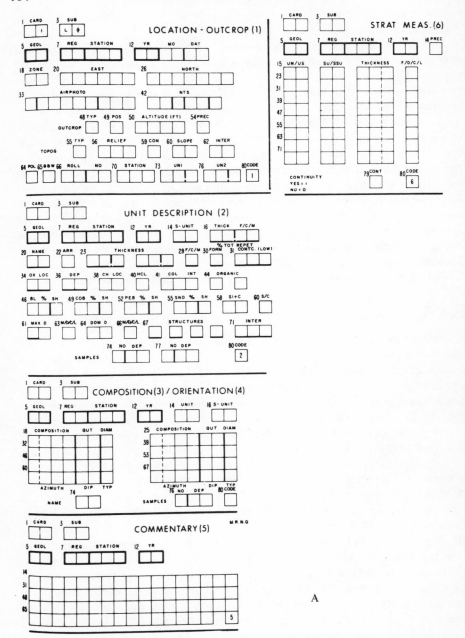

8.11 LEDA stratigraphic coding system: (A) field data forms; (B) application to a cor

GEOLOGICAL MAPPING BY COMPUTER 185

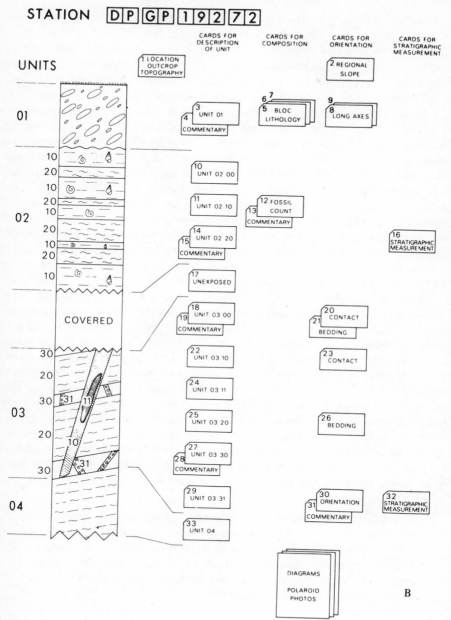

ical section (David and Lebuis, 1976; reproduced by permission of Pergamon Press Ltd)

Display and analysis are clearly subsequent to and dependent on retrieval of the relevant subset of the whole data set (although the form of the retrieval may be affected by the intended use), and some of the analysis or display capabilities may already be part of the data base management system. We are thus left with problems of storage and problems of retrieval. De Heer and Bie (1975) describe some of their requirements for a suitable data system. It should accept input in the form 'attribute = value' since the skills required to convert ordinary field information into fixed length codes is rarely available. It should allow hierarchical relationships between data elements. It should allow retrieval by Boolean expressions (for example, if clay is less than 10 per cent). Finally, the data base itself should recognize four main types of data sets (points, lines, areas, and descriptional data sets independent of geographical coordinates) and any relationships between them. Many industrial systems cannot handle these requirements. Many, geared to textual data, are unable to retrieve all data elements whose values of a particular attribute fall within a numeric range. Even more commonly, the relationships between points, lines, and areas on a map are not readily definable using only the hierarchical relationships available with most commercial systems, since these particular relationships are best expressed as ring structures.

Other required operations are more readily available with many systems. These include the storage, updating, and rearrangement (editing, sorting, merging, subfile creation, etc.) categories mentioned earlier. Although special difficulties may arise in particular applications, these more properly belong in the domain of computing science and will not be described further here. It is this writer's opinion, however, that until spatial coordinates and relationships are recognized as separate and distinct from other data types, and are stored and treated according to their own rules, truly meaningful and conceptually elegant geological data base management systems will not be achieved. It is not, for example, easily possible to extract all water samples collected within a particular drainage basin unless such a relationship was specifically defined when the water sample data were entered. Nevertheless, such an operation is easily done by humans using a map, and questions of this form are basic to geological thinking. Hutchison (1975, p. 5) states that:

> statistical and other analyses ... can be fed back, as part of an iteration, to allow evaluation of domain boundaries. In this way data [points] may be given geological significance and as domain boundaries are changed, then so must the geological context of all contained data. Many geologists do this every day and it is essential that any computer system for field geology must have the same capacity.

The previously mentioned work on triangulations of a plane is a tentative step in this direction.

Considering the relationship between points, lines, and areas, Hutchison again comments that geological field data are frequently considered as points, as if they had been acquired by a blind machine. He continues:

> The geologist does not stop at one point A, look down at his feet and record all significant data within a radius of three metres of his feet, then close his eyes and walk blindly onwards for a hundred metres or one kilometre and then stop, look down, open his eyes at point B and record all data within a radius of three metres. Instead he is more concerned in the first instance in establishing the relationship between data set A and data set B, and recognizing (using his own inboard computer) whether or not there is a difference, and if so, what its nature might be. Each hour and each day in the field is spent working essentially interactively with the rock patterns to build up a picture of the field setting. ... I would contend ... that there has been a general failure to recognize that geological contacts are prime data located on geological field maps ... whose nature results from observations *between* data set A and data set B. On this basis, 'spot' data within units are therefore of lower rank than the data for contacts between units (Hutchison, 1975, p. 3).

So much for the spatial aspect of the data set. Sutterlin and May contend that the slow acceptance of data base management systems by geologists is because:

> They are accustomed to dealing with naturally occurring objects and phenomena which are quite complex and only partially understood. As a result, the data about these objects and phenomena are varied and complexly interrelated—that is, complexly structured—and the earlier techniques and concepts designed for more predictable and less highly structured data of the commercial environment were less than completely successful when applied to the management of geological data (Sutterlin and May, 1978, p. 30).

They comment that modern computing advances have largely eliminated the need for data coding, although it is often used in the field to obtain a measure of uniformity and completeness at the data collection stage. While much work has been done on geological data structures, little of it is directly relevant to mapping technology exclusively. Further information may be found in Sutterlin *et al.* (1974) and Dickie and Williams (1975).

CONCLUSIONS

We have discussed some of the major types of geological map and how various workers are attempting to automate different procedures. We can summarize by saying that contour type map production by computer is alive, well, and hopefully moving towards second-generation techniques. These will probably involve taking the topology or neighbourhood relationships of data points into account. Data transformation (often statistical) prior to display is fairly widely used, but again spatial relationships are not normally considered. In lithology mapping, topology is being seriously examined by a few workers, but automation is rare beyond selective data point posting. This last, along with the necessary field data acquisition and management, is progressing steadily.

Where contouring is not involved, the use or non-use of computers for map production seems mainly to depend on whether automated statistical or other data transformations have already been performed. If so, or if field coding forms were used, automated posting of data values will probably be used. In geology, computer-generated point, line, and area symbolism has not been extensively examined, although some consideration has been given to the problems of the computer-assisted drawing of boundaries between groups of similar data points. A particularly bright spot is the recent concern over the topological relationships between neighbouring geological entities (such as rock units) on a map. Nevertheless, with this exception (and that of contouring) computer usage in geological map production is primarily in data transformation rather than display.

REFERENCES

Agterberg, F. P. (1974). *Geomathematics*. Elsevier, Amsterdam.

Basham, P. W., Forsyth, D. A., and Wetmiller, R. J. (1977). 'The siesmicity of northern Canada', *Can. J. Earth Sci.*, **14**, 1646–1667.

Berner, H., Ekstrom, T., Lilljequist, R., Stephansson, O., and Wikstrom, A. (1975). 'GEOMAP'. In W. W. Hutchison (ed.), *Computer-based Systems for Geological Field Data*. Geol. Surv. Can. Paper 74–63.

Bouillé, F. (1976). 'Graph theory and digitization of geological maps', *Math. Geol.*, **8**, 375–393.

Bouillé, F. (1977). 'Structuring cartographic data and spatial processes with the Hypergraph-Based Data Structure. In G. Dutton (ed.), *First International Advanced Study Symposium on Topological Data Structures for Geopraphic Information Systems*, vol. 5. Harvard Laboratory for Computer Graphics and Spatial Analysis.

Bouillé, F. (1978). 'Survey of the HBDS applications to cartography and mapping'. Presented paper at Harvard Computer Graphics Week, Cambridge, Mass.

Burk, C. F., Jr. (1973). 'Computer-based storage and retrieval of geoscience information: bibliography 1970–72', Geol. Surv. Can. Paper 73–14.

Burns, K. L. (1975). 'Analysis of geological events', *Math. Geol.*, **7**, 295–321.

Charlesworth, H. A. K., Langenberg, C. W., and Ramsden, J. (1976). Determining axes, axial planes and sections of macroscopic folds using computer-based methods. *Can. J. Earth Sci.*, **13**, 54–65.

Dahlberg, E. C. (1975). 'Relative effectiveness of geologists and computers in mapping potential hydrocarbon and exploration targets', *Math. Geol.*, **7**, 373–394.

David, P. P., and Lebuis, J. (1976). 'LEDA: a flexible codification system for computer based files of geological field data', *Computers and Geosciences*, **1**, 265–278.

Davis, J. C. (1973). *Statistics and Data Analysis in Geology*. John Wiley & Sons, New York.

Dawson, E., and Newitt, L. R. (1977). 'An analytic representation of the geomagnetic field in Canada for 1975. Part I: The main field', *Can. J. Earth Sci.*, **14**, 477–487.

de Heer, T., and Bie, S. W. (1975). Particular requirements for the Dutch WIA earth science information system. In W. W. Hutchison (ed.), *Computer-based Systems for Geological Field Data*. Geol. Surv. Can. Paper 74–63, pp. 78–81.

Dickie, G. J., and Williams, G. D. (1975). 'Development of a computer-based file on oil and gas pools', Geol. Surv. Can. Paper 75–22.

Forgotson, J. M., Jr. (1960). 'Review and classification of quantitative mapping techniques', *Bull. Amer. Assoc. Petrol. Geol.*, **44**, 83–100.

Garrett, R. G. (1974). 'Field data acquisition methods for applied geochemical surveys at the geological survey of Canada', Geol. Surv. Can. Paper 74–52.

Gold, C. M. (1977). 'The practical generations and use of geographic triangular element data structures'. In G. Dutton (ed.), *First International Advanced Study Symposium on Topological Data Structures for Geographic Information Systems*, vol. 5. Harvard Laboratory for Computer Graphics and Spatial Analysis.

Gold, C. M., and Kilby, W. E. (1978). 'A case study of coal resource evaluation in the Canadian Rockies using digital terrain models'. Presented paper at Harvard Computer Graphics Week, Cambridge, Mass.

Gold, C. M., and Maydell, U. M. (1978). 'Triangulation and spatial ordering in computer cartography'. Presented paper, Can. Cartog. Assoc. Third Annual Meeting, Vancouver.

Gold, C. M., Ramsden, J., and Charters, T. D. (1977). 'Automated contour mapping using triangular element data structures and an interpolant over each irregular triangular domain', *Computer Graphics*, **11**, 170–175.

Govett, G. J. S., Goodfellow, W. D., Chapman, R. P., and Chork, C. Y. (1975). Exploration geochemistry—distribution of elements and recognition of anomalies. *Math. Geol.*, **7**, 415–446.

Harbaugh, J. W., and Merriam, D. E. (1968). *Computer Applications in Stratigraphic Analysis*. John Wiley & Sons, New York.

Howells, K., and McKay, A. G. (1977). 'Seismic profiling in Miramichi Bay, New Brunswick', *Can. J. Earth Sci.*, **14**, 2909–2977.

Hrŭska, J., and Burk, C. F., Jr. (1971). 'Computer based storage and retrieval of geoscience information: bibliography 1946–69', *Geol. Surv. Can. Paper* 71–40.

Hutchison, W. W. (1975). 'Introduction to geological field data systems and generalized geological data management systems'. In W. W. Hutchison (ed.), *Computer-based Systems for Geological Field Data*. Geol. Surv. Can. Paper 74–63, pp. 1–6.

Hutchison, W. W., and Roddick, J. A. (1975). 'Sub-area retrieval system (SARS) used on the Coast Mountains project of the Geological Survey of Canada'. In W. W. Hutchison (ed.), *Computer-based Systems for Geological Field Data*. Geol. Surv. Can. Paper 74–63, pp. 32–38.

Jaquet, J. M., Froidevaux, R., and Vernet, J. P. (1975). 'Comparison of automatic classification methods applied to lake geochemical samples', *Math. Geol.*, **7**, 237–266.

Journel, A. (1975). 'Geological reconnaissance exploitation—a decade of applied statistics', *Can. Inst. Mines Bull.*, **68**(758), 75–84.

Kearey, P. (1977). 'A gravity survey of the central Labrador Trough, northern Quebec', *Can. J. Earth Sci.*, **14**, 45–55.

Knuth, D. E. (1968). *The Art of Computer Programming*, vol. 1. Addison-Wesley, Reading, Mass.

Langenberg, C. W., Rondell, H. E., and Charlesworth, H. A. K. (1977). 'A structural study of the Belgian Ardennes with sections constructed using computer-based methods', *Geol. en Mijnbouw*, **56**, 145–154.

Males, R. M. (1977). 'ADAPT—A spatial data structure for use with planning and design models'. In G. Dutton (ed.), *First International Advanced Study Symposium on Topological Data Structures for Geographic Information Systems*, vol. 3. Harvard Laboratory for Computer Graphics and Spatial Analysis.

McCammon, R. B. (1968). 'The dendrograph: a new tool for correlation', *Geol. Soc. Amer. Bull.*, **79**, 1663–1670.

Nichol, I., Garrett, R. G., and Webb, J. S. (1969). 'The role of some statistical and mathematical methods in the interpretation of regional geochemical data', *Econ. Geol.*, **64**, 204–220.

Parker, M. A. (1946). 'Use of International Business Machine technique in tabulating drilling data', *Illinois State Acad. Sci. Trans.*, **39**, 92–93.

Peucker, T. K. (1972). 'Computer cartography', Commission on College Cartography Resource Paper no. 17. Assoc. Amer. Geographers, Washington, D.C.

Ramsden, J. (1975). 'Numerical methods in fabric analysis', Ph.D. Thesis, University of Alberta, Edmonton, Canada.

Ramsden, J., and Cruden, D. M. (in press). 'Estimating densities in contoured orientation diagrams', *Geol. Soc. Amer. Bull.*

Robinson, J. E., and Charlesworth, H. A. K. (1975). 'Relation of topography and structure in south-central Alberta', *Math. Geol.*, **7**, 81–87.

Smith, W. (1815). *A Delineation of the Strata of England and Wales with Part of Scotland.* S. Gosnell, London.

Sutterlin, D. G., Aaltonen, R. A., and Cooper, M. A. (1974). Some considerations in management of computer-processable files of geological data', *Math. Geol.*, **6**, 291–310.

Sutterlin, P. G., and May, R. W. (1978). 'Geology: data and information management'. In J. Belzer, A. G. Holzman, and A. Kent (eds.), *Encyclopedia of Computer Science and Technology*, **9**, 27–56. Marcel Dekker Inc., New York.

Wishart, D. (1975). *CLUSTAN 1C User Manual.* University College, London.

The Computer in Contemporary Cartography
Edited by D. R. F. Taylor
© 1980 John Wiley & Sons Ltd

Chapter 9
Census Mapping by Computer*

Frederick R. Broome and Sidney W. Witiuk

INTRODUCTION

Background

Maps have traditionally played several useful roles in the collection and dissemination of census data. Until quite recently, almost all of these maps were produced using the traditional techniques associated with manual cartography. Starting from standard *topographic maps* (see Chapter 5) for rural areas or from recompiled base documents in urban areas, cartographic draftsmen working in census agencies around the world have long been producing *field maps* to assist the enumeration process. In many instances this geographic framework would then be transcribed and generalized as small-scale *index maps* for inclusion with statistical tables. In certain publications, dating back at least as far as the 1870 Decennial US Census, a limited number of *statistical maps* have been manually prepared to portray the spatial distribution of census data.

Computers have been used for more than three decades to store, edit, retrieve, and tabulate vast quantities of statistics on socio-economic data collected during the census. Indeed the Bureau of the Census was one of the prime movers of computerization as evidenced by the creation of the Hollerith punch card by Herman Hollerith, then a U.S. Census employee. However, except for some data compilation and classification procedures, and a few early isolated attempts at developing primitive statistical mapping capacities, the computer has not until very recently been used on a large scale as an aid in the preparation of various types of maps for census purposes.

Computer processing power was never the issue; even the earliest computers available to statistical agencies and other users of census data were quite capable of processing data in cartographic form.

Historically, the unfulfilled challenge was to produce cartographic products of sufficiently high quality at sufficiently low costs. Due to decreased costs and increased precision of specialized computer peripherals for encoding and displaying

---

* The views expressed are those of the authors and not of the agencies in which they are employed.

graphics (see Chapter 4), this challenge is now within reach for a large number of applications.

This chapter highlights some of the more notable successful applications of automated and semi-automated statistical and non-statistical mapping within and/or in concert with census agencies around the world.

Structure and Definition

The applications considered within this chapter are organized according to the following framework shown in Table 9.1. While straightforward in nature, this categorization has been used by the authors solely to structure the myriad applications of computer-assisted cartography (C.A.C.) into manageable groups for analysis and comparison. A number of geocartographic systems have been developed which are capable of producing both statistical and non-statistical maps at quality levels suitable for both internal purposes and for external distribution, and these systems can be applied either by the bureaus themselves or by other agencies under contract. In such cases the systems will be dealt with in different ways several times, and their scope and value will become evident in terms familiar to cartographers and managers of cartographic facilities.

A number of terms and concepts may be new to non-technical readers. To facilitate communication a brief but important analysis of the components of a 'typical' system for C.A.C. is presented, in which definitions of selected terms have been incorporated.

Table 9.1

Map producer	Map Type			
	Non-Statistical		Statistical	
Mapping by census bureaus	Internal needs	External needs	Internal needs	External needs
Other agencies	Internal needs	External needs	Internal needs	External needs

Elements of Typical Systems for C.A.C.

In order to provide a common framework for analysing the applications considered in this chapter, the following model of the components of a typical computerized mapping system is presented (Figure 9.1).

'On-line' or 'batch' systems for the computer-assisted production of statistical and non-statistical maps typically require the input of three distinct types of information:

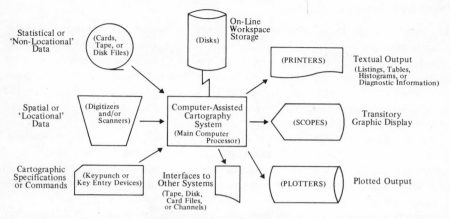

9.1 Elements of a typical computer-assisted cartography system (Statistics Canada, 1972b)

(i) Non-locational statistical data, for example average income per person;
(ii) location or 'spatial' data, for example, census tract boundaries;
(iii) cartographic specifications or 'commands', for example, scale, rotation, titles, mapping technique, etc.

In the case of 'batch' systems all of these inputs must be provided in machine-processable form prior to submitting a 'job' for execution, for example, preparing a deck of cards to control map preparation. 'On-line' systems typically provide at least some degree of interaction, experimentation, and/or editing during execution.

Locational data are usually geographically encoded ('geocoded') in relation to some standard reference frame for example, latitude/longitude or U.T.M.) by use of manually controlled digitizing devices, semi-automated line-following systems, or by fully automated raster scanning devices. Fuller details of these techniques have been given by Boyle in Chapter 4.

Output Data Formats

In addition to the intended point, line, area, and text artwork produced for temporal (for example on C.R.T. scopes) or permanent display (for example on plotters), most *geocartographic* systems also provide or permit standard 'printer output' of tables, histograms, and diagnostic information. The term 'geocartographic' is used here to denote the perhaps subtle difference between a system restricted to the encoding, manipulating, and decoding of solely cartographic features and a system that is also able to deal effectively with statistical data sets and hierarchies of geographic data.

A limited number of the applications to be discussed within the framework of this chapter involve geocartographic systems that have the added characteristic of being able to produce results on a variety of different types of output devices. These systems are termed 'device independent'.

Processing Elements

The logic of a geocartographic system is contained within one or more computer programs which are read into the hardware's central processing unit (C.P.U.) and associated memory. The way in which this is handled by computer scientists varies from application to application and usually depends upon tradeoffs between computer and program size, speed, and the importance of flexibility in the sequencing of work elements. Programs that are able to execute on a variety of different types of hardware are called 'machine-independent'.

Because large amounts of locational and/or non-locational data are typically or potentially involved, it is usually necessary to restructure the data and to save it on relatively fast access, 'on-line' storage devices such as a data disk or drum. This restructuring and relocation of the input data is undertaken to reduce both the space required for storing the data and the elapsed time for re-accessing it.

Finally, a number of geocartographic programs are able to 'link' to other ubiquitous packages such as SPSS (Statistical Package for the Social Sciences) and SYMAP via standardized interfaces.

Thus, the most popular and widespread geocartographic systems are those that are: low cost, device-independent, machine-independent, able to provide both on-line and batch processing, and already contain standardized interfaces to other widely used packages. In Chapter 10, Waugh discusses these design features in greater detail and documents the way in which one geocartographic system, GIMMS, has developed to meet these objectives.

COMPUTER-ASSISTED MAPPING BY CENSUS BUREAUS

Applications of C.A.C. *within* Census bureaus will focus primarily upon the North American experience. Proximity to many of the centres of some of the fastest technological development during the 1950s, 1960s and early 1970s, and the vast expanses of territory to be mapped, are two probable causes for the high degree of initiative and independence of the Canadian and U.S. Census Bureaus in the field of C.A.C.

Census-Produced Non-Statistical Maps for Internal Needs

The U.S. Bureau of the Census must be credited with some of the most extensive early efforts in the area of computer-assisted, non-statistical map production in census bureaus around the world. In support of the decision to use a mail-out/mail-back methodology for enumerating large urban areas (Standard Metropolitan Statistical Areas, SMSAs), it was necessary to geocode extensive and consistent lists of valid household addresses. In response to the quality assurance component of this undertaking, the U.S. Bureau of the Census developed the GBF/DIME (Geographic

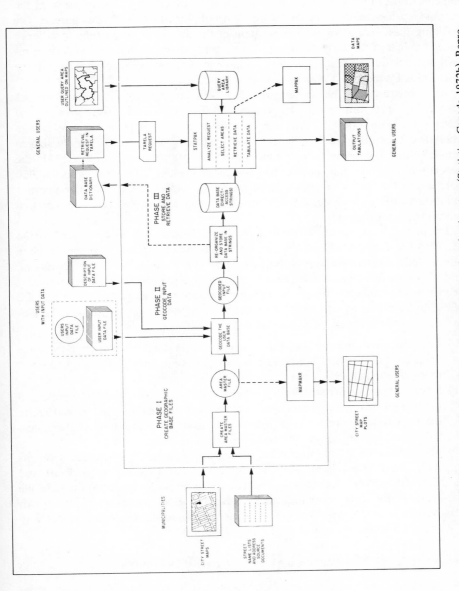

9.2 Flowchart of the geographically referenced data storage and retrieval system (Statistics Canada 1972b). Reproduced by permission of the Minister of Supply and Services Canada

Base File/Dual Independent Map Encoding) system and methodology. While the actual production of these files was decentralized to the various SMSAs, the primary impetus, most of the funding, and a substantial involvement in quality assuring the graphic representation of the files were contributed by the U.S. Census Bureau. The resulting files have a potential application of the linkage of files of the local level. Indeed, much of the early digitizing of street and other important network features was carried out on the digitizing equipment designed and constructed in the mid-1960s by the U.S. Census.

At approximately the same time the Canadian Census was launching its 'geocoding' program using the Geographically-Referenced Data Storage and Retrieval System (GRDSR) primarily to respond to a need for a micro-level data analysis capacity (Figure 9.2). Starting with geocartographic bases for only sixteen municipalities (in fourteen centres of population 100,000 or more) in 1971, the geocoding program has grown to 133 municipalities (100 centres of 50,000 population) for the 1976 census. Unlike the U.S. experience, the file preparations associated with census geocoding activities in Canada to date have been concentrated within the central agency.

The plot of the Lemoyne urban area master file (Figure 9.3) was produced by the MAPMAKR sub-system and demonstrates that while the base contains a large number of important features (street pattern, rivers, municipal limits, street names, and addresses), the level of detail clearly indicates that the bases were constructed for primarily internal needs. However, once the very extensive and often costly exercise of building a digital geocartographic model of the urban region is complete, the files can be used for a variety of cartographic purposes:

(i) The Canadian Bureau has tested and is evaluating prototype systems for producing field maps with double-line street casings from the single-line street representations in the data base as shown on Figure 9.4;
(ii) Since census data are structurally related to the geographic features represented in these geocartographic data bases, both agencies have the ability to restructure the data base for regional or network analysis as a result of mathematical 'grouping' operations on the cartographic entities (Figure 9.5);
(iii) The proliferation of interactive topographic mapping systems such as AUTOMAP (Canadian), M&S, SYNACOM and CALMA (American) and CLUMIS (British) provides an opportunity to vertically integrate the base file creation for field mapping, index mapping and statistical mapping. An evaluation of the cost effectiveness of such an approach is currently under way in the Canadian Bureau of the Census (Figure 9.6).

Census-Produced Non-Statistical Maps for External Distribution

Most users of census-type data in tabular form have long appreciated the occasional index map demonstrating the geographic framework underlying a given table or set of charts. Until effective digitizing systems became available and until the recent

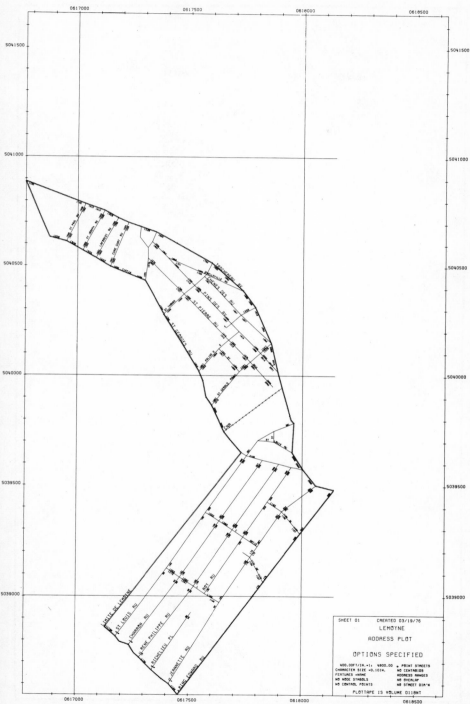

9.3 A.M.F. plot of Lemoyne urban area. Reproduced by permission of the Minister of Supply and Services Canada

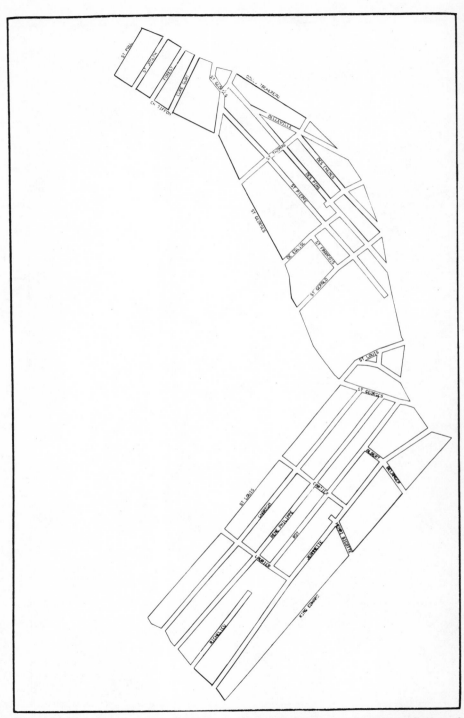

9.4 Double line plot of Lemoyne urban area. Reproduced by permission of the Minister of Supply and Services Canada

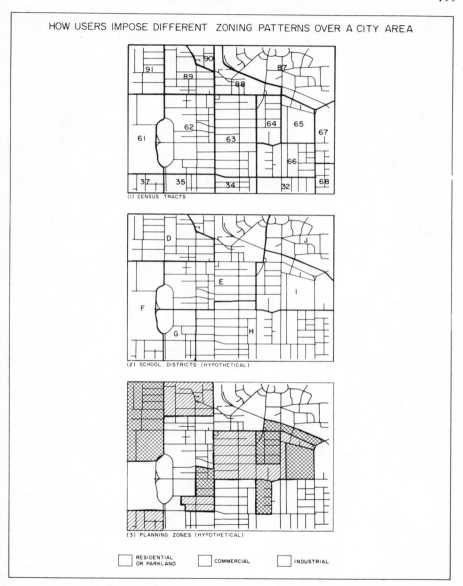

9.5 Different zoning patterns for a city area restructuring the data base for regional network analysis (Statistics Canada, 1972b). Reproduced by permission of the Minister of Supply and Services Canada

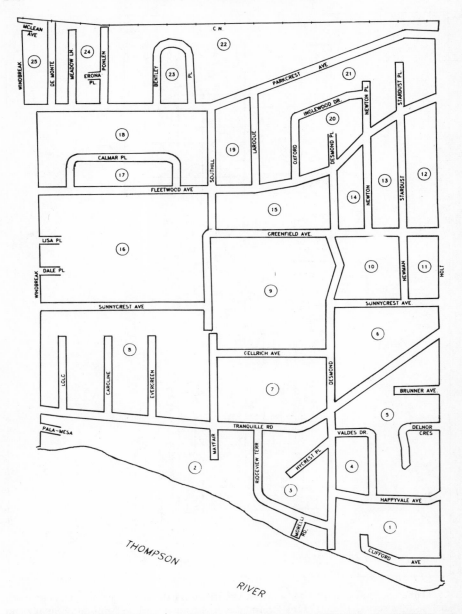

9.6 Prototype field collection map—Kamloops CT-16 (using AUTOMAP, SYNTHESIS and PLOT (Statistics Canada, 1978). Reproduced by permission of the Minister of Supply and Services Canada

interest in having these reference frames in machine-processable form (for example in the semi-automated production of thematic maps or cartograms such as the Isodemographic map of the U.S.A.—Figure 9.7) the impetus for the computer-assisted production of index maps has been minimal.

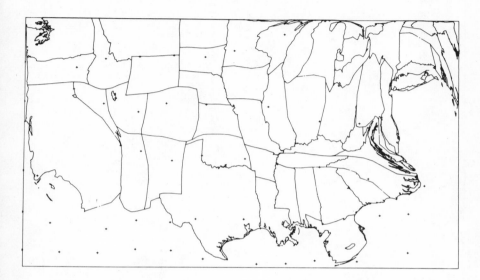

9.7 Isodemographic map of the U.S.A. (Tobler, 1976). Reproduced by permission of Waldo Tobler

Figure 9.8, from the bulletin released in 1978 by Statistics Canada entitled, *Agriculture Graphic Presentation*, shows not only that index maps can be produced from base files intended for thematic mapping purposes, but also that other semi-automated techniques such as photocomposition can be incorporated within the production process.

The possibly non-aesthetic reference map presented earlier as Figure 9.4 is distributed to external users of census data to satisfy a number of practical applications. Such maps are used to delineate and manually 'geocode' study area boundaries such as traffic or planning zones for focusing retrieval of socio-economic data. They serve as a valuable adjunct to coding procedures associated with applications dealing with manual or semi-automated linking of address data. Another, and final example, is the use of the geographic structure and locational attributes (especially the U.T.M. coordinate values) of these reference documents as the framework for geographically referencing the incidences of spatial phenomena. A typical application is the coding of traffic accidents to street intersection node numbers. In such cases the graphic and the digital representation of the map are usually sent to the user.

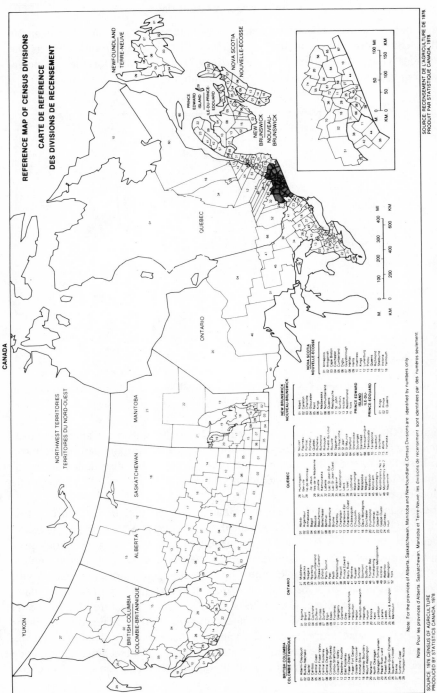

9.8 Computer-assisted production of index maps (GIMMS System, Statistics Canada, 1978). Reproduced by permission of the Minister of Supply and Services Canada

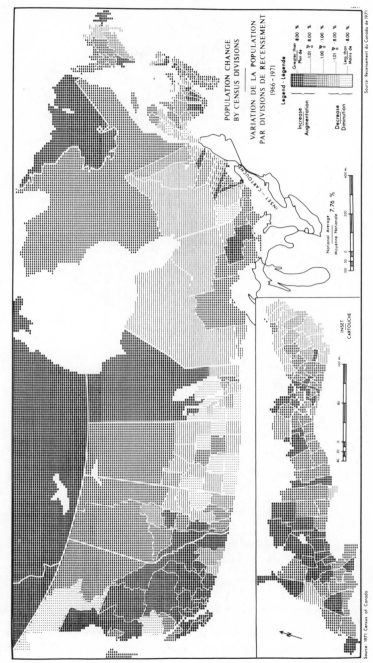

9.9 An example of the use of SYMAP (Statistics Canada, Census Data Display Unit). Reproduced by permission of the Minister of Supply and Services Canada

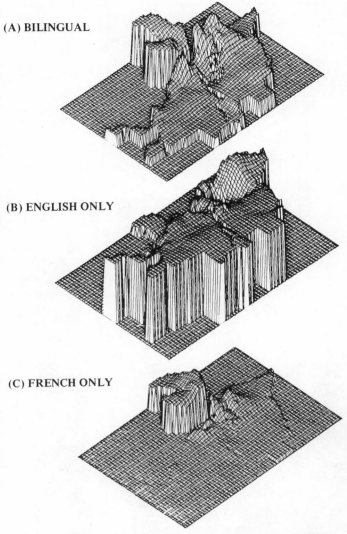

9.10 An example of the use of PREVU (these are three-dimensional representations of 'language surfaces' in Ottawa–Hull (Taylor, 1977))

Census Produced Statistical Maps for Internal Needs

Starting with batch systems such as SYMAP (Figure 9.9) distributed by the Laboratory of Computer Graphics and Spatial Analysis, Harvard University) and PREVU (Figure 9.10) (developed at Carleton University), medium- and low-quality choropleth, proximal, and isopleth maps and statistical surface representations have found increasing use and acceptance.

The more recent systems such as PRISM and CALFORM (distributed by Harvard), WISMAP (University of Wisconsin), GIMMS (University of Edinburgh) and PILLAR (University of Ottawa), to name a few, have provided geocartographic capacities of increasing resolution and complexity.

A number of factors have recently combined to encourage the use of computer produced *statistical* maps for internal analytical purposes:

(i) the conversion to E.D.P. and data base technology by statistics gathering agencies;
(ii) the rapidly growing volumes of data being collected by such agencies;
(iii) the rising costs of manpower involved in the data preparation and analysis process and the need to streamline these processes;
(iv) the decreasing costs of specialized graphics hardware peripherals—especially cathode ray tubes;
(v) a growing awareness and familiarity by social and economic scientists of the analytical techniques and methodology of geocartographics.

Perhaps the most important of these factors is the last one mentioned, since only through time has the sound application of these techniques been clearly delineated. Early misuse of spatial association, as demonstrated on a map, to infer causal relationships amongst subsets was quickly unmasked by quantitative tests for statistical significance. In the meantime, however, a number of legitimate applications have evolved and are becoming increasingly entrenched as analytical aids.

At the lowest level of analytical sophistication, computer-assisted statistical maps allow the analyst to identify the absence or unreasonableness of data and thereby assist with the internal quality assurance function.

At the global level, associations such as typically *urban*, *rural*, *cultural*, *economic*, or *regional* help bring into greater focus potential relationships worthy of more rigorous or more detailed analysis. From a marketing viewpoint the statistical maps not only serve as a data reduction instrument but are also useful in selecting the 'more interesting' distributions. Finally, the computer-produced map (especially the 'continuous choropleth' and the isopleth maps) are particularly useful in generating regionalizations.

Both the Canadian and the U.S. Bureaus of the Census have used digitally encoded index values to *dasymmetrically* 'weight' the distribution of the mapped statistical values, thereby bringing greater meaning to the presentation. Figure 9.11 shows how the statistical data were concentrated within census divisions according to the measure of improved land in relation to the amount of land in agriculture.

Finally, a potential application only now under consideration is the production of statistical maps in support of data gathering activities. Maps showing response rates,

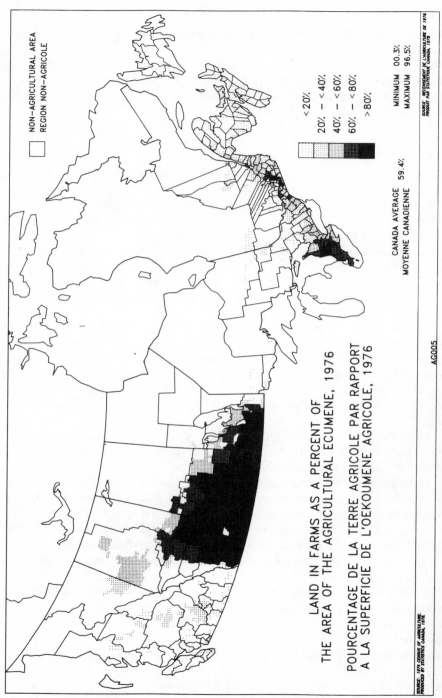

9.11 Dasymmetrically weighted choropleth mapping (GIMMS and C.G.I.S. systems, Statistics Canada, 1978). Reproduced by permission of the Minister of Supply and Services Canada

linguistic character, or mode of enumeration, can be used as an aid to logistics control by data gathering divisions.

Census-Produced Statistical Maps for External Distribution

By far and away the most visible and most impressive activity involving computer-assisted mapping within census bureaus is the production of statistical maps for external distribution. To date, five types of presentation have been undertaken between the two North American Censuses:

(i) Single volumes summarizing national level activity on a particular sphere of the economy, for example, agriculture. The 1969 (215 maps) and 1974 (307 maps) Agricultural Graphic Summaries by the U.S. Bureau of the Census and the Canadian 1976 (114 maps) Agricultural Graphics Presentation are three of the best-known applications of computer-assisted statistical mapping produced internally by census bureaus. Figure 9.12 shows examples from these publications.

(ii) Series of volumes, one for each urban region, showing comparable data. To be consistent, the details of the U.S. Urban Atlas Series for the 1970 Census must be discussed in a later section as the work was subcontracted to an external agency. The Canadian Census has produced, using 1976 data, a prototype for one centre (Ottawa) that is modelled, in part, on this earlier undertaking. Figure 9.13 demonstrates the use of GIMMS (see Chapter 10) to produce camera-ready black and white renditions of census tract data in a substantially smaller format.

(iii) Single or multiple maps on single sheets to highlight selected variables for a given publication. Two examples of this type of application are seven statistical maps (one dot map and six choropleth maps in colour) included in the 1976 Canadian Census Bureau's *Status* bulletin.

(iv) Maps 'made to measure', on demand for users of census data. This type of service has been available on a limited basis in Canada since the mid-1970s, first using SYMAP and more recently employing the GIMMS package, and is becoming a popular adjunct to statistical tables. Applications in Canada have ranged from urban area market analysis studies by commercial outlets, through provincial level analysis of inventories of forest stocks to national level studies of health and unemployment data. The 'one of a kind' nature of these types of request, in contrast to the tens or hundreds of maps in an atlas or graphics summary, places heavy demands upon the design and dependability of the chosen computer-assisted mapping system. These issues are discussed in Chapter 10.

(v) Providing direct access to a data base of public domain data via 'user-friendly' geocartographics systems. The White House initiated N.A.S.A./Census Domestic Information Display System (DIDS) is a prototype development that has shown the technical feasibility and utility of such systems. When considered

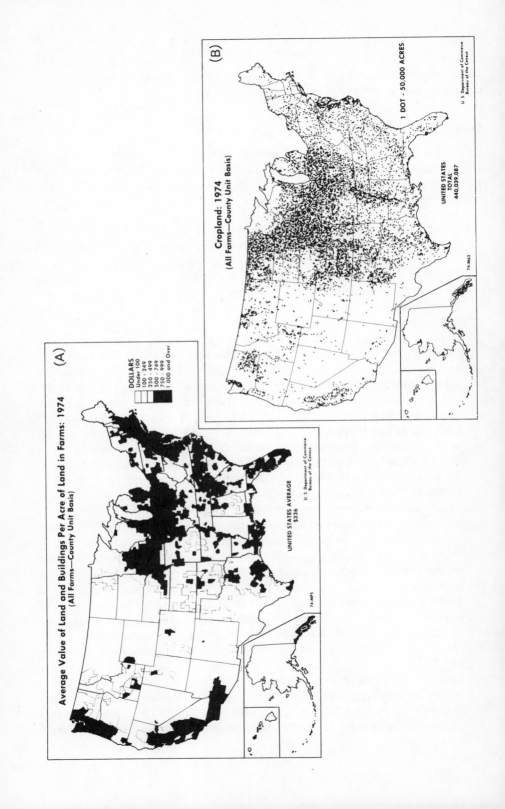

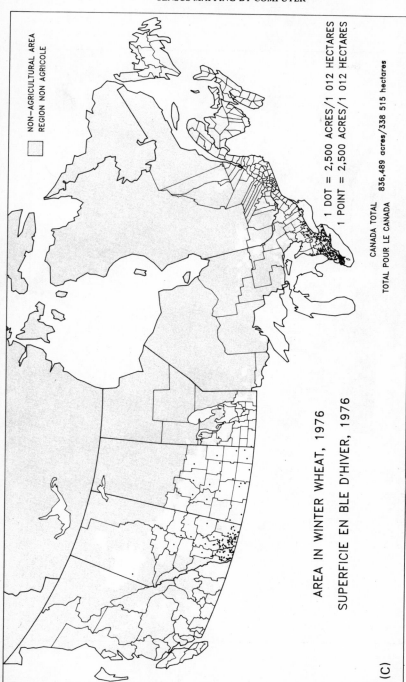

9.12 Geocartographic summaries of national level distributions. Maps A and B: U.S. Bureau of the Census, 1978; Map C: Statistics Canada, 1978. Reproduced by permission of the Minister of Supply and Services Canada

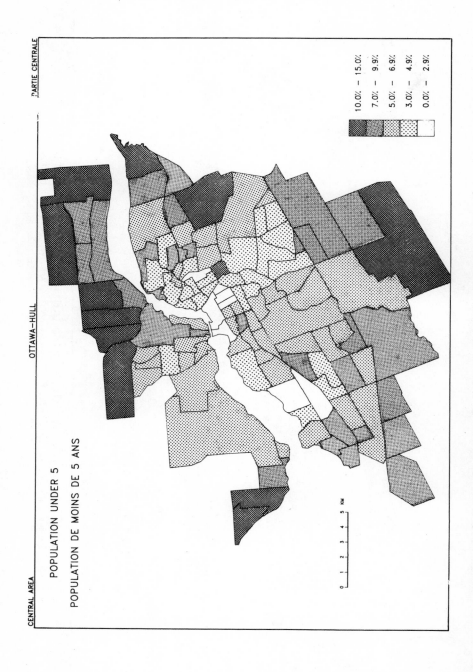

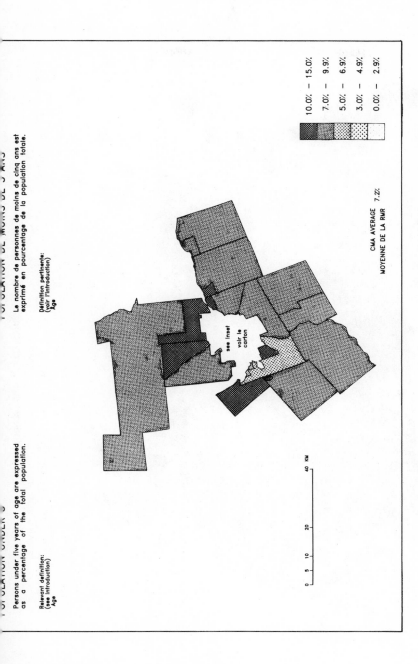

9.13 Two examples of the use of GIMMS to plot census tract data (Statistics Canada, 1978). Reproduced by permission of the Minister of Supply and Services Canada

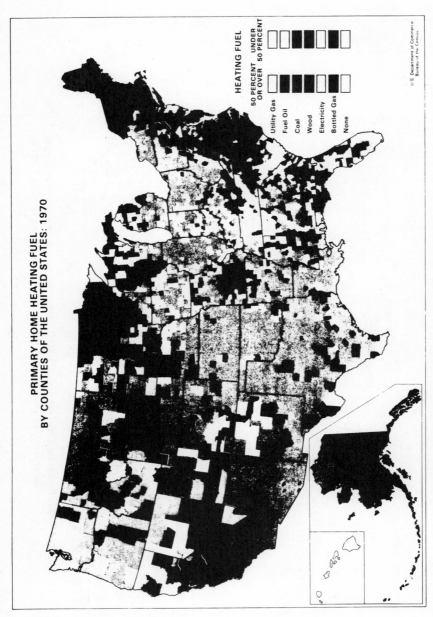

9.14 The Domestic Information Display System (an example of 'hands on' computer-assisted cartography, U.S. Bureau of the Census and NASA, 1978, which can now be produced interactively using the DIDS system)

in the light of developments such as PRESTEL and TELIDON, in Great Britain and Canada respectively, the impact of such technology on reducing the volume of *hard copy* statistics and manually produced high resolution maps of the type in national atlases may be quite dramatic (Figure 9.14).

COMPUTER-ASSISTED MAPPING BY OTHER AGENCIES

Whereas a large percentage of the computer-assisted mapping applications *vis à vis* Census requirements are conducted 'in-house' in North America, the opposite is the case in Europe. This is not to say that North American universities, commercial establishments, and local, regional, province/state and national agencies are not also processing census data by means of computer-assisted geocartographic systems to meet their specific requirements. The major difference lies in the fact that, outside North America, internal census requirements for C.A.C. products are typically (at the present time) fulfilled by external agencies. The constantly lowering costs and increasing flexibility and reliability of these systems may well result in the more widespread distribution in the next few years.

Non-Statistical Maps for Internal Needs

Topographic mapping at the national level was discussed by Harris in Chapter 5. To date, little has been achieved in providing cartographic bases in a digital format suitable for internal census purposes. Given the massive undertaking associated with national level topographic data bases and the different requirements (cartographic versus geocartographic data structures) it is unlikely that progress will be rapid. A notable exception to this pessimistic outlook is the ongoing exchange of reference bases between the Canadian Census and the Canada Geographic Information System of the Department of the Environment.

The rapid growth of regional and urban information systems based on geocartographic data bases offers greater hope for the near future. The need to take advantage of data from local and regional systems such as LAMIS, RGU, and TRAMS is far less in Europe with its longstanding tradition of extensive and large-scale mapping programs.

Thus the linkage of census systems to regional and local geocartographic data bases is likely to proceed more rapidly in the younger and more extensive area of North America. Exclusive of the decentralized GBF/DIME program, the only exchange of this type known to the authors is a direct update interface between Statistics Canada's GRDSR and the Infor-Estrie system in Sherbrooke, Quebec.

Externally Produced Census Maps for External Distribution

One of the best examples of this type of mapping is related to legislative requirements for the production of maps showing municipal boundaries or federal electoral districts. In Canada, such maps are produced by the Surveys and Mapping Branch of the Department of Energy, Mines, and Resources, and are based, in part, on census field collection maps.

Externally Produced Statistical Maps for Internal Census Use

Maps are seldom produced externally for solely internal purposes. Frequently, contractual relationships are established for maps intended for external distribution and as part of these undertakings a series of lower quality/lower cost maps are produced to quality-assure the data and to finalize the selection of themes.

Externally Produced Statistical Maps for External Distribution

Census data and therefore maps of census data are not solely processed and distributed by Census organizations. Commmercial interests such as information service bureaus (for example, COMPUSEARCH, *Financial Post*, etc.) and planning consultancies use census and other data extensively in the provision of their services. As evidenced by the 1978 'Graphics Week' sponsored by Harvard University, the commercial sector agencies are increasing their use of computer-assisted mapping systems. It is also notable that systems that produce lower-quality maps (usually requiring manual touch-ups) are being replaced by systems able to produce camera-ready maps of medium quality.

The focus of this section, however, will be on work conducted externally to the census agency but considered part of the census program. While there are examples of this type of 'sub-contracting' in North America (for example, The US Census Urban Atlas Series was produced at the Lawrence Berkley Laboratory), in Europe it tends to be the rule rather than the exception.

Figure 9.15 exemplifies the many 'grid square' maps produced for the British Census by the Census Research Unit at the University of Durham. In Britain and France a large percentage of the statistical mapping activity is conducted by the Department of the Environment and l'Institut Geographique National (I.G.N.) respectively.

The Department of the Environment has been involved with computer-assisted mapping since the mid-1960s through its sponsorship of the development of the LINMAP and COLMAP. Currently they are producing on the order of 700 maps per year by computer-assisted means. The I.G.N. converted to semi-automated systems more recently but has more than made up for its late start with an impressive array of map types at medium and high quality.

With a growing number of universities offering training in C.A.C. (especially to geographers and planners) an increasing involvement and volume of production of statistical maps is being initiated by planning professionals in regional and local government. The growth and spread of associations such as the Urban and Regional Information Systems Association (URISA) to Britain (BURISA) and Australia

9.15 Grid square map (Census Research Unit, University of Durham)

(AURISA) and the Segment Oriented Referencing Systems Association (SORSA) is promoting the global transfer of this technology.

CONCLUDING REMARKS

The increasing use of general E.D.P. technology for collecting, processing and distributing census information over the past decade has given great impetus to the successful application of computer-assisted map production systems, both within and external to census bureaus, to meet internal and dissemination needs.

The optimism and enthusiasm of the late 1960s has given way to the more sober and experienced perspective of the late 1970s. The costly process of building geocartographic data bases to support the detailed geographic structuring of census information is now realized to be only practical in the context of increased use of these files for computer-assisted production products such as *field* maps, *index* maps and *statistical* maps. The combination of increased reliability and reduced costs of specialized hardware for cartography, the more general availability and stability of software packages for statistical and non-statistical mapping of medium to high quality, and the emergence of multi-disciplinary specialists in 'geocartographics' are leading to the widespread adoption of this technology by other government agencies (local, regional, and national), universities, and more recently, the commercial sector. Having resolved many of the early conceptual and implementation difficulties of building geographic base files and of producing low- to medium-quality statistical maps, census bureaus involved in computer-assisted mapping now seem to be focusing on ways of improving the quality of these files and systems to increase their general utility and on finding ways of reducing their cost.

While it is not possible to say that all of the technical problems are currently solved, especially as relates to the input of large volumes of data, the greatest impediment to greater acceptance of geocartographics' technology is not technological in nature. Nor is the area of costs and benefits of primary concern, although the more narrow the objective the more difficult the problem of amortizing initial data base creation costs. While a 'dirty word' to many professionals in the field of geocartographics, the major difficulty currently is the *marketing* of the products. Having survived the difficulties of the first generation of geocartographic systems, many managers of E.D.P.-oriented processes are far more sceptical about the short- and medium-term benefits of increased technology. This, in some ways, healthy scepticism is particularly noticeable within census bureaus where the issue of *risk* must play as important a role as *cost*, given the nature of the census-taking activity. This perhaps explains why many of the selected examples of developments within census bureaus tend to be described as prototypes. Were it not for the fact that census bureaus, like most other government departments around the world, are expected to be more effective and responsive to the ongoing needs of their users, the 5–10-year cycle of the census would have a tremendous impact on the rate of internal development of geocartographic services. Fortunately these systems are being increasingly viewed as useful tools for improving the appeal and effectiveness of the census product line and developers are being encouraged to meet these needs, albeit on a cost-recovery basis.

However, if the marketing aspect of geocartographic systems development is given proper attention, the prospects for the future are most encouraging. Seemingly unrelated activities such as the formation of professional societies to share technology and develop teaching curricula; experiments in the use of graphics by young children to solve advanced problems in set and group theory; studies in how maps are perceived and the proper use of colour; standardization of terms and development of geocartographic data transfer standards; special symposia of leading international experts to document the limits of the state of the art; and support of high profile demonstration projects are all combining to lay the framework for continued success.

REFERENCES

Statistics Canada (1972a). *1971 Census of Canada Advance Bulletin*, Catalogue 92–753 (AP-2), Ministry of Supply and Services, Ottawa.

Statistics Canada (1972b). *GRDSR: Facts by small areas*, Ministry of Supply and Services, Ottawa.

Statistics Canada (1978). *1976 Census of Canada, Agriculture, Graphics Presentation*, Catalogue 96–871, Ministry of Supply and Services, Ottawa.

Taylor, D. R. F. (1977). 'Graphic Perception of Language in Ottawa-Hull', *The Canadian Cartographer*, **14**, 21–34.

Tobler, W. (1976). 'Analytical Cartography' *American Geographer*, **31**, 18–26.

United States Bureau of the Census (1969). *Census of Agriculture 1969* Volume V, Special Reports, Part 15, Graphic Summary, United States Department of Commerce, Washington, D.C.

United States Bureau of the Census (1978). *Census of Agriculture, 1978*, Volume IV, Special Reports, Part 1, Graphic Summary, United States Department of Commerce, Washington, D.C.

United States Bureau of the Census, and N.A.S.A. (1978). *Domestic Information Display Demonstration* (Unpublished pamphlet).

Witiuk, S. W. *et al.* (1978). 'On the feasibility of automating certain aspects of the production of Census reference maps—Progress Report No. 1' Uncatalogued internal document, Census Field, Statistics Canada, Ottawa.

The Computer in Contemporary Cartography
Edited by D. R. F. Taylor
© 1980 John Wiley & Sons Ltd

Chapter 10

The Development of the Gimms Computer Mapping System

Thomas C. Waugh

INTRODUCTION

In the introductory chapter of this book, Taylor made a distinction between 'automated mapping' and 'computer mapping'. The production of thematic maps by computer was, in fact, one of the earliest applications of computer assisted cartography (C.A.C.), but it is only recently that systems producing high-quality thematic maps have emerged. This chapter will concern itself with the development of one such system—GIMMS. Although GIMMS belongs to the 'computer mapping' end of Taylor's spectrum, it produces much higher quality graphics than earlier thematic mapping programs such as SYMAP (Fisher, 1966).

GIMMS (Waugh, 1979; Richer, 1978; Waugh and Taylor, 1976) is a large computer program, developed over several years, which is primarily used for the production of medium- to high-quality maps. Figure 10.1 shows a simple example produced by the system on a relatively inexpensive drum plotter. The system is used by many organizations, primarily academic or governmental. It is used for research purposes to produce 'quick-look' maps and as a system to produce high-quality maps for publication, notably by Statistics Canada.

The system has a wide range of cartographic options (see Figure 10.2) but most users do not know (or need to know) all of the possible options; the system will use sensible defaults where appropriate. The production of simple maps is therefore relatively easy. The number of parameters that may be set in the total system, however, exceeds 700 and the utilization of these allows the production of very complex maps.

The system has developed over most of the last 10 years and has undergone much change and enhancement since then. This chapter will trace the development of the system and will focus on some of the problems encountered and decisions taken during that development.

CHRONOLOGY OF DEVELOPMENT

The system was originally started while the author was a visiting student at the

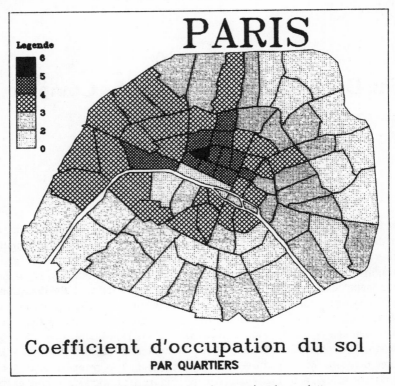

10.1 A GIMMS plot on an inexpensive drum plotter

Laboratory for Computer Graphics and Spatial Analysis at Harvard University during 1969–70. A previous system, CMS (Chloropleth Mapping System) (Waugh, 1972), developed at the University of Edinburgh by the author, was recognized at that time to be a limited system in several ways. It was geared to lineprinter output, as is SYMAP, and therefore was extremely limited in cartographic terms. At that time, it could only deal with area descriptions (that is, no point or line descriptions) and numeric statistical data. In addition, the computer design had two major flaws: the system was written in a local (Edinburgh) language, IMP, similar to ALGOL, and all the data values for all the area zones were held in memory all the time; this meant that the system was quite large when running, especially on significant amounts of data. CMS did, however, have many good points and continued to be developed (in parallel to GIMMS) for several years.

It was intended to develop a new system which would be extremely comprehensive, contain the best points of CMS and other systems, and be methodologically sound both in computing application system terms as well as geographic system terms.

The first two modules (IOPACK and DATMAN) were developed with assistance

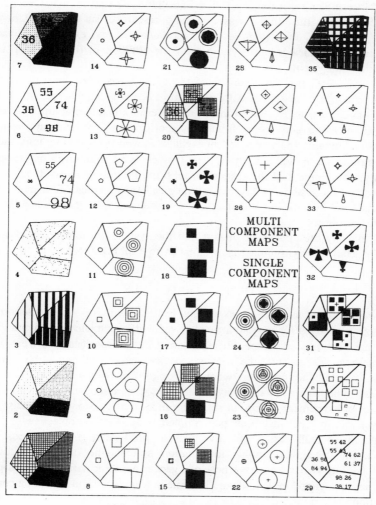

10.2 GIMMS symbolism chart

from the Department of Landscape Architecture at Harvard University and were solely a general-purpose manipulation package for statistical data. The DATMAN module has undergone little change in the 10 years since that time but will be completely replaced by mid-1980.

The GIMMS system, as a recognizable system, was first developed with funding from Leyland Systems Inc., of Boston, to provide a vehicle for the topological encoding of street networks. However, it was never used for this purpose and the system returned to Edinburgh in 1970 where it went through sporadic development during which its image changed to that of a mapping system. During the period 1973 to 1976 it was being distributed to other installations on an informal basis.

In 1976, while the author was a visiting researcher at Carleton University, Ottawa, the system underwent several major changes and expanded its facilities dramatically as a cartographic system with the addition of the *COMPILE map compilation subsystem.

Since late 1976, the development of the system has been funded to provide facilities which are missing and to make the system more general. Most of this funding has come from Statistics Canada who were using it to map the 1976 census. Development is currently being funded by the Scottish Development Department who wish to enhance the geoprocessing facilities as well as the cartographic facilities.

Throughout the period since 1971, the University of Edinburgh has supported, indirectly, development of the system through the Program Library Unit.

The cost of the system in effort has been considerable. At least four man-years of effort has gone into the development of the system as well as at least £10,000 of computer resources.

This diversity of support has meant that the system has undergone changes of image in detail but has retained its original cartographic outlook.

DESIGN OBJECTIVES

It is always difficult to recollect accurately, in the absence of documentation, what the design objectives for a system were. Even with complete documentation, the author has encountered a case where apparent design specifications were, in fact, implementation results since the original specifications proved impossible to meet. It is useful to catalogue, however, what the design objectives appeared to be and to investigate whether or not they have been met. A later section deals with areas of change or enhancement which, in some instances, reflect poor or imprecise design objectives and in other instances, poor implementation of the objectives.

While the overall design objective was to produce a general-purpose, user-oriented, integrated, geographic information system, the design objectives, in general, fall into two categories—those of a geocartographic nature (geocartographic is used here as defined by Witiuk and Broome in Chapter 9) and those of a computer application system nature.

Geocartographic Objectives

The geocartographic objectives were of a general nature but the major ones were that the system should be general-purpose, it should produce better graphics (for example, maps) than were available, and it should have the capability for extensive manipulations, especially of a geographic nature.

The requirement that the system be general-purpose was paramount. It should be able to handle point, line, and area geographic descriptions, and associate both numeric and alphabetic data with the descriptions. GIMMS will, in fact, do this but with restrictions. The data types are, in general, well separated. The system does not allow data files of mixed types and does not allow, at present, alphanumeric data to be included on the same file as geographic descriptions. The separation has had the effect of placing the system, at the moment, firmly at the thematic end of processing systems since data from the base mapping systems as described by Harris in Chapter 5, have a mixture of geographic descriptions (coordinates) and alphanumeric data (feature codes). One of the results of the separation of the geographic data from the alphanumeric data is that it has been difficult to deal with missing data. The data input procedures have required that data are available for every zone described although there is the capability to subset the descriptions. Thus, although the system is a general-purpose one in a broad sense, there are restrictions which are sometimes of an irksome nature.

The system was intended to be methodologically sound and is, in fact, still one of the few systems that reads descriptions of areas by segments and then topologically checks the linkage. This has the effect of ensuring that geographic area descriptions are complete and correct and will not cause the system to produce unexpected results due to faulty descriptions. For example, the results of a point-in-polygon operation are undefined if the polygon is not a correctly closed polygon. This design objective has therefore been fairly well achieved.

The system indubitably produces better graphics than lineprinter systems although it does not at present produce isoline maps. In all other areas, however, the graphics from GIMMS are superior to lineprinter systems and many other plotter systems. It is probably the only system which produces three different generic shading styles—cross-hatch, dot screen, and bar shading (the development version now also includes half-tone and symbol shading). It can also produce dot maps, single and multicomponent point symbol maps, and rudimentary line symbolism.

The design objective to ensure a capability of extensive manipulation has remained more or less as an objective. Although the capability exists, GIMMS at present does little in the way of geographic processing. This will be discussed later in this chapter. GIMMS has extensive manipulation capability in a statistical sense (for example, calculation of new variables) but the development effort has gone into the mapping subsystem, not into the geoprocessing functions.

Thus, the geocartographic design objectives have, in general, been met albeit with some restrictions.

Computer System Design Objectives

The design objectives for the computer system were much more specific than the geocartographic objectives and much more difficult to follow. The major objectives were that the system should be user-oriented, integrated (yet modular), have a simple (for the user) job control interface, have a simple graphic interface, be usable in a batch or interactive environment, and have no arbitrary restrictions. While these are still the objectives, the fervour to stick to them has been watered down a little.

The problem of the graphic interface is described in a later section, as is the problem of user-orientation—two of the most demanding and difficult objectives. The most demanding objective, however, is that the system should have no arbitrary restrictions; all that can be hoped and planned for is to minimize the effect of them. Earlier revisions of GIMMS had a plethora of restrictions, most of which have since been removed, or are in the process of being removed. The removal of one such restriction is described in the section on 'Major System Changes or Additions' in this chapter. Many restrictions still exist, however, such as a limit of 20 variables to be mapped with a multicomponent symbol.

During the development of GIMMS, the ideal of an integrated but modular system has been modified slightly so that the integration is more important. The system is highly modular internally but it is only possible for certain areas of the source code to be used external to the system. It was originally expected that the system would be broken into several programs but the availability of overlay facilities has meant that this has not been necessary and complete integration means better and more efficient use of space and facilities.

The intention to have a simple job control interface for the user has been achieved. All user-defined GIMMS files have the same format and this has been true since 1970. A major change, however, is in the offing.

One criterion which has never been a design objective should perhaps be mentioned, and that is the running cost of the system. No real attempt has been made to make the system really efficient. There have been two major reasons for this: first, it has long been the author's belief that computing costs would continuously drop; and secondly, that it was more important to spend effort on providing new facilities than making the system more efficient. The former has proven to be true, with the possible exception of plotting costs which in some installations remain artificially high. In any event, the system is not particularly expensive to run. A GIMMS map is about six to ten times more expensive than a lineprinter map of the same distribution but has increased precision, a resolution at least 20 times better than a printer, and a better product. A recent map of 75 shaded zones cost approximately £4.

INITIAL IMPLEMENTATION AND SUBSEQUENT DEVELOPMENT

The initial implementation of GIMMS occurred during the period November 1969 to July 1970 and consisted of six modules—an input system for segment data, a simple editor for the data, a topological checker for the segment data, a polygon linker for the segments, the DATMAN data manipulation system, and the mapping module which would only shade simple areas.

Apart from the shading mechanism, most graphics were done by the graphic library used (for example the CALCOMP basic software) through a fairly crude graphic interface. It can thus be seen that most of the effort was put into the data collection and manipulation features and the cartographic facilities were fairly crude.

As it was intended that the system should be run on-line, four of the modules were not only designed to be interactive but also conversational. That is, the system prompted the user for all his input as though conversing with him. As subsequently became clear, these were the very modules which would normally be run in batch mode and this made the giving of the input data more difficult.

From 1970 to 1976 most of the effort was spent on improving the mapping system. During this period, it was expanded to include point and label mapping, multicomponent point mapping, and basic line mapping. There were experiments with anaglypth mapping which were not successful, and colour offset maps, which were. Various new input systems were added to cope with the new data types.

In 1976, while a visiting researcher at Carleton University, the author took the opportunity to rethink some major parts of the system and to create new modules to solve basic problems. Foremost among these was the development of the GPIS system (Waugh, 1977) which replaced the old command system, and the introduction of the *COMPILE subsystem which allowed the interactive design of maps. In a sense, one led to the other. It was clear that a *COMPILE system was required but the existing language could not support such a system. Thus the GPIS was born. It has since been developed to be a very effective man–machine communication tool and is the mainstay of most of the commands.

It was during this period also that the graphic interface was improved and almost all the graphics were organized to be generated by GIMMS including all scaling, windowing, and text production.

Since 1976 there have been major reconstructions of the system such as the replacement of the *BOUNDCHECK and *AREAFILE modules with the *CHECK/*POLYGON module, the addition of dot mapping, the additions of enhanced text capabilities, and the general cleaning up of some commands.

It can be seen that the development of the system has meant that its image has changed to primarily that of a thematic mapping system rather than the original image of a data collection and checking system. Some of the more important changes or enhancements are described in the following two sections.

MAJOR SYSTEM CHANGES OR ADDITIONS

There have been several major changes in GIMMS since its initial implementation and these have had a profound effect on the way that a user approaches the system. Some of them have been in the system as a system and others have been of a cartographic nature. This section looks at some of the major changes or enhancements made to the system which are not particularly cartographic in nature.

Changes occur for several reasons in GIMMS. The major reason is that the previous method has either become very difficult to use or is not capable of expansion to meet new needs. It is impossible to design from scratch a system which will be capable of doing all things for all users on all machines. It is, however, possible to try to minimize the changes necessary. Some of the changes reflect either poor original design or a change in understanding of the problems involved. Some reflect user feedback, particularly in the context of 'bottlenecks' where users were finding certain things difficult to do.

In any change, one of the design priorities was that the new system should be as upward-compatible as possible. That is, that user commands given under the old system should be valid, if at all possible, under the new system. Files should also be upward-compatible.

Graphic Interface

The original ideal in the development of GIMMS was that the graphics whould be device- and graphic-library-independent. This ideal has remained a constant problem ever since. The first implementation separated GIMMS and the graphic libraries by making sure that there was no physical connection. GIMMS produced a text file of plot commands, and after the GIMMS run was finished, a simple program to decode and draw the file was run.

This method proved unsuitable for several reasons. First, it was clumsy as one had to run two programs to get a plot. Secondly, it produced a large file of plot commands, and thirdly, it militated against interactive graphics.

It was decided, therefore, to include calls to graphic libraries directly from GIMMS. Unfortunately, there are many graphic libraries, and at the time of writing, the system will call the GINO-F, CALCOMP, GERBER, IG, and ERCC packages. In fact, since almost every installation modifies the CALCOMP package, it currently can call four different versions of the CALCOMP library. In addition, it can still produce its file of plot commands.

Current work has changed the method yet again. The as-yet unreleased development version calls a standard set of routines and it is the role of the installation to write these routines to interface to their graphic package. In fact there are two distinct interfaces: one to a general purpose plotter and one to an interactive device. An interface to the Tektronix 4000 series storage display tubes is provided for the latter interface, and interfaces to GINO-F, CALCOMP, GERBER and other graphic packages for the former.

This method has distinct advantages over previous methods: it is more easily implemented for a new package and is more easily documented; it is simpler to include, or ignore, sophisticated facilities such as could be provided by intelligent terminals; and it removes the requirement for the user to specify which graphic library is to be used—all that is required is to specify which device is required. For example, *PLOTPARM PLOTTER selects the plotter and *PLOTPARM T4010 selects a Tektronix 4010 display.

User Interface

The user interface, that is, how the user tells the system what to do, is critical to the effective use of GIMMS. The original GIMMS language was an outgrowth of the

CMS (Waugh, 1972) language. However, as GIMMS grew in size and complexity the language, such as it was, was not really able to cope. This was primarily because the user commands consisted basically of positional parameters of various types. While parameters could be bypassed if a different type of parameter was specified, if they were all of the same type then the original parameters had to be given before specifying the new ones. For example, if a text string was to be oriented at an angle other than a horizontal one, then the size of the text needed to be specified, even if it was the default. Thus, *TEXT 5.4 7.2 0.5 45 'text string' was required even though the 0.5 (height of text) was the default assumption. As the number of options grew, this method became very unwieldy. The *TEXT command alone has 33 values that may be specified and the new (as yet unreleased) *LEGEND command has more than 150 values.

It was necessary to change the language, and the opportunity was taken to greatly enhance the facilities. The resulting language is more fully described in Waugh (1977) but the basis of the new language is that it is keyword oriented allowing the user to specify parameters in any order.

Thus:

*TEXT POSITION = 5.4, 7.2, ANGLE = 45, TEXT = 'text string'

would have the same effect as the previous example. Since, however, the keywords are ordered, then the command as given previously will also work, as will:

*TEXT 5.4 7.2 ANGLE 45 'text string'

and many other variations. This reduced confusion which existed in the previous system.

The new system (called the GPIS) has many more facilities such as the capability to allow parameters to be set by a graphic cursor. This latter requirement was prompted by the development of the *COMPILE system. In addition, the system provides facilities to 'help' the user by providing him with a list of options and their type while working interactively. The GPIS is easier to use, understand, implement, and document than the previous system and is easier to expand to include new facilities.

The *COMPILE Subsystem

The development of the GPIS was prompted by two major factors: first, the previous method had become clumsy, on which many users had commented, and secondly, it was not interactive and could not support interactive graphics. The latter requirement came from the recognition of a user bottleneck.

When using lineprinter mapping systems, the design of the map is not too important since it is going to be fairly crude in any case. On a line plotter, however, the design of the map is quite important. It is obvious, and jarring to the eye, if a piece of text overlaps another, or the map proper. On most systems there are two ways to circumvent this problem. The first is to submit several runs, one after the other, to correct errors. This may take a long time. There are few installations which provide faster than an overnight turnaround time for graphic plots (although the situation is improving). This may mean that it may take 3 or more days for three or more runs to provide a 'correct' plot. This is extremely wasteful of time and money. The second method is to 'preview' a plot on a graphic display if one is available. This speeds the plotting process but may still be time consuming.

The *COMPILE subsystem was developed to remove this bottleneck and operates by providing an interactive graphic map design tool. The user sits at a graphic display terminal and draws the elements of a map (outlines, text, legend, etc.). He may redraw each element with different parameters until he is satisfied and then 'keep' that element as an 'object'. A collection of 'objects' defines a map skeleton, and the commands necessary to produce these objects may be printed by the system and used to create a final map either interactively or, more usually, in batch.

This subsystem allows the user to design and test out graphically a map skeleton in a matter of minutes rather than the days often required by batch means, and thus reduces one of the common bottlenecks found in computer-mapping systems.

Data Bottleneck

Due to an original design error there was, for many years, a limitation on the maximum coordinate value allowed during the polygon creation subsystem in GIMMS. This limit was 32767 which will be immediately recognizable to many persons who have worked with computers. The value was, effectively, the upper limit of any coordinate system to be used in the system. This proved to be unacceptable as a reasonable resolution, on the ground, of a national mapping system of even such a small country as the U.K. (National Grid) produced coordinate values in excess of this limit.

The polygon creation system was therefore replaced by the new *POLYGON module which was a vast improvement over the previous method. This new module also provided the topological checking function of the system, replacing the *BOUNDCHECK module with the *CHECK module which provides much better error and diagnostic messages.

MAJOR ENHANCEMENTS OF A CARTOGRAPHIC NATURE

The development of GIMMS in a cartographic sense has probably benefited most from user feedback since it is the requirements of the users and how the system matches these requirements that measure the quality of a system. Since the results are visual, then mismatching of the facilities to the requirements is often very obvious. The developer of GIMMS has therefore put a high priority on meeting user requirements.

In addition to meeting these requirements, there has been a conscious attempt to improve the cartographic quality of GIMMS output, as opposed to providing a wider range of facilities at lower quality.

Over the years a considerable amount of work has been done to improve the cartographic quality of GIMMS, some of the changes being small, others requiring a large amount of work. The enhancements described below are not a complete set of enhancements but indicate those types of areas which must be considered and dealt with by any person developing a system that is claimed to be cartographic. It is no longer acceptable to excuse poor quality because it is computer produced!

Double Lines

One of the problems encountered in using many plotters, especially inexpensive drum type plotters, is that there is a certain amount of backlash (or slackness) in one or both of the driving motors for x and y movements. If a series of adjoining polygons is then drawn, the common boundaries appear as a double line. This is due to the fact that if two adjoining polygons are drawn in the same direction (for example, clockwise) then the common boundary will have been drawn in two different directions, once for each polygon, and the backlash on the plotter drive causes the same nominal point to appear in a slightly different location. The same effect causes 'tramlining' in shading cross-hatch patterns not parallel to the axes.

If all that is available to the system is a polygon description, then the problem of double lines is virtually unsurmountable. Fortunately, GIMMS carries segment information into its polygon files and therefore it is possible to ascertain, when drawing a particular polygon, which parts of the polygon have already been drawn as part of another polygon. These parts are then omitted. A decision on preserving information 4 years earlier paid off in the solution of this problem.

Facilities Provided by the Graphic Interface—Text and Dashed/Dotted Lines

The original implementation of GIMMS did not directly produce text or dotted or dashed lines. It called the appropriate subroutines in the installation-provided graphic software.

While most graphic libraries provide text drawing facilities, not all (for example, CALCOMP basic software) provide dotted/dashed line generation. Therefore in 1977 dotted, dashed, and chained line generation was added to GIMMS to operate at the lowest level so that all features of GIMMS could be drawn using dotted and dashed lines.

The example of text production is more important, however. Originally, GIMMS called the appropriate text routines in the graphic library supplied by the installation, to draw text characters. This had several drawbacks! First, the method of text generation varied tremendously among different graphic libraries and even in the same library. For example, five installations which mounted GIMMS found that their computer services had modified the CALCOMP basic software routine calls to produce text to suit their local environments. Secondly, the text produced was not of good quality and varied among different libraries. The CALCOMP software insists on drawing a line through the zero number to distinguish it from the letter 'O'. Thirdly, the graphic libraries, in general, did not provide many 'esoteric' facilities such as superscript, subscript, italic, varying boldness of characters, and so on.

In 1976, therefore, the opportunity was taken to implement a text system within GIMMS containing all of the above-mentioned features and some others. This made GIMMS independent from the graphic libraries for text production. An example of GIMMS text can be seen in Figure 10.3. During 1976 and 1977 it became obvious, partly from user feedback, that the flexibility of text production was an important part of an automated mapping system. Therefore in 1977 major enhancements were made to the text subsystem. The most important was probably the addition of the 'Hershey' alphabets (Wolcott and Hilsenrath, 1976) also shown in Figure 10.3. These were restructured into a more appropriate form and a general-purpose alphabet input system was developed so that the user could specify his own alphabets. In addition, the text facilities were further enhanced to include such facilities as centring of multiple line strings, right justification of blocks of text, and variable ratio text.

GIMMS basic alphabet

This is a Hershey font

THIS IS A BLOCK ALPHABET

10.3   Types of text available in GIMMS

Certainly the provision of at least the 'Hershey' alphabets must be considered a necessity for any computer mapping. It would be nice to think, however, that any computer mapping system should be able to provide most, if not all, of the current GIMMS facilities such as the ability to produce italicized script as written by a left-handed person! The GIMMS text facilities have been used external to GIMMS for other uses. The system (without the Hersey alphabets) is used as the text-generation system of the ODYSSEY system produced at Harvard University's Laboratory for Computer Graphics and Spatial Analysis, and the GIMMS version of the Hershey alphabets is used in the system developed by M. McCullagh at the Universities of Nottingham and Kansas. Since GIMMS alphabets conform to a standard format, it is hoped that sharing of new alphabets will occur over the next few years.

Area Shading

One of the biggest drawbacks to most vector plotter graphics systems has been the relative crudeness of the shading produced. The normal method is to produce cross-hatched lines for shading. While suitable for the darker end of the grey spectrum, especially for solid black, it is not appropriate for the lighter end of that spectrum since the texture is crude, which makes it confusing to the eye and difficult to identify.

Originally GIMMS only produced cross-hatch patterns, but with the use of dotted and dashed lines for the patterns, various effects could be achieved. Since, however, the positioning of the dots and dashes along a line was independent of any others along another line, the effects were somewhat random.

The opportunity was taken to rewrite the shading algorithm and it was decided to extend the capabilities of the system to include three basic types of shading: cross-hatch, dot screen, and bar shading (shown in Figures 10.2). These can be mixed in any order and provide a much wider variety of shading. With a little bit of user-effort, true half-tone shading can also be produced.

Recent work has introduced true half-tone shading automatically, the provision of continuous shading (Tobler, 1973; Brassel and Utano, 1978), and the facility to allow user provided symbols for the shading pattern.

It is of interest to note why the shading algorithm was rewritten. The previous system was a polygon-by-polygon shading system (as are most, if not all other systems). This was found to have a major problem when attempting to produce high-quality maps using ballpoint or liquid-ink pens. The effect is most noticeable with felt-tip pens (which would not, however, be used for high-quality work). When the pen is dropped at the edge of the polygon and drawing commences across the polygon (a solid line is assumed), the ink flow cannot keep up with the demand. Thus near the middle and end of a line, the line density is less than at the beginning. This has the effect of producing a band of darker ink around the polygon boundaries. Even slowing the plotter down does not always eliminate this problem. The effect is particularly noticeable when adjoining polygons have the same shading patterns and can be quite disastrous if solid black shading (which turns out to be an even solid black) is being drawn for colour offsets.

As part of the rewrite of the shading algorithm, the *JOINSCAN command was implemented. This has had the effect of not shading polygon by polygon but saving the shading lines until the whole map is complete. The shading is then drawn, but shading lines which are co-linear are combined to form a single line. For example, if five adjoining polygons have the same shading line pass through them, then only one line will be drawn instead of five. The effect is thus minimized and solid shading occurs only at the peripheries of the shaded block.

Two interesting spin-offs have occurred from this development. First, the effect of doing a *JOINSCAN produces faster plotting since the stopping and starting motions are reduced. This is particularly noticeable on a graphic display screen such as Tektronix since the device can draw a long line as fast as a short one in most circumstances. There can be up to a ten-fold increase in plotting speed on such a screen. The second spin-off is that under certain circumstances the processing time to draw the map may be significantly reduced.

The *COMPILE Subsystem

The *COMPILE subsystem, briefly described previously, does more than reduce a bottleneck as described above. It provides a genuinely new computer cartographic tool allowing a cartographer to interactively design a map. While it is true that it reduced a bottleneck, the effect of the bottleneck was not just that designing a map took longer, but that the effect was quite often that the map was only half designed. This is quite an important problem with general-purpose computer-mapping systems. It is not that the system cannot do what it is required to do, but that the user (who may not have much cartographic pride) will not expend the effort to produce a 'good' map. Using the *COMPILE system, which allows the user to manipulate the design of a map, means that the map is more likely to be designed rather than just thrown together.

Systems such as this allow designing in a way which is more appropriate to the cartographer and is nearer to more traditional methods of map design. It is confidently expected that a large and growing proportion of published thematic maps will be designed this way in the next few years.

CURRENT DEVELOPMENT

The GIMMS system is under continuous development. At the time of writing (January 1979), the system as distributed (Release 3) has already been superseded by the development version in many ways, both in cartographic terms and in system terms.

Modifications have streamlined the system, made it smaller, and cleaned up some of the older subsystems. In addition, many of the older commands have been converted to the GPIS system to bring them into line with the newer commands.

The graphic interface has been completely rewritten to make it easier to implement and maintain, and to prepare for the expected growth of intelligent graphic terminals. A new data transfer and diagnostic aid has been provided.

In cartographic terms there have been many notable changes, mostly prompted by user feedback, particularly from (and funded by) Statistics Canada. The graphic interface now supports 'shielding' as well as 'windowing' which makes the production of insets, in particular, much easier. The position automatically chosen by GIMMS to be the centroid of a polygon was not always satisfactory and sometimes (to the cartographer's great consternation) appeared outside the polygon. A new facility to move unsatisfactory centroids by pointing on a screen if desired was developed.

A completely new *LEGEND system was developed which provides much more flexibility in the creation of shaded point, and dot legends. This was in direct response to the suggestion of a cartographer (with whom the author does not agree) that shaded legends should always have the lightest shade at the top of the legend which at that time was done with difficulty by GIMMS. The new command allows the easy creation, possibly unique in a general-purpose system, of being able to produce a density legend for a dot map.

Ever since the original development of GIMMS, the user has had the option of choosing equal class intervals or of specifying his own. Recent developments mean that the user has the option of twelve different methods to choose class intervals and also the option of ignoring classes and producing continuous symbolism for shaded and point symbols.

With these and many other enhancements, the system will more easily be able to provide a high-quality cartographic product, and it is hoped that cartographers will continue to influence the development of the system so as to meet their needs.

FUTURE DEVELOPMENTS

There are various directions planned for GIMMS developments in the foreseeable future. Many will be complete by mid-1980, funded by the Scottish Development Department.

A new direction for GIMMS is the movement into non-cartographic graphics. In conjunction with Statistics Canada, facilities to produce graphs, histograms, bar charts, pie diagrams, and the like are being developed. These may, of course, be used in conjunction with maps and will provide a powerful tool to the cartographer to add supplementary information to maps.

The DATMAN module was the first GIMMS module to be developed and it is planned to replace it with a much more flexible system which will also allow grid systems in preparation for isopleth mapping and three-dimensionsl mapping.

New cartographic facilities are planned such as the addition of line (or flow) symbolism and user-defined symbol mapping.

In many cases, a particular user will develop skeletons of commands that he uses many times with only a few changes each time. It is planned to develop a macro facility to allow a user to create common groups of commands which can then be modified and executed by one command. This facility will make it easier for teaching purposes, since students may be taught the use of macros rather than the original commands.

There will also be a move towards more general-purpose geoprocessing with the ability to deal with more topographic type data and selection and search facilities will be added. Disaggregation and aggregation facilities are also planned.

These and other developments will make the system very powerful and will no doubt lead (probably with user feedback) to further developments which in turn will lead to even further developments.

CONCLUSION

This chapter has attempted to show some of the problems and decisions involved in the development of the GIMMS system. As the section on future developments indicates, the system is under constant development and it is expected that this will continue for some time.

The system is, at the moment, one of the foremost computer cartographic systems in the world and is one of the first to move computer cartography from the 'isn't it interesting what computers can do—but it's not cartography' era into the 'produced by computer—I don't believe it!' era; as such it heralds the mass introduction of such tools for practical map-making in many different fields.

REFERENCES

Brassel, C. E., and Utano, J. J. (1978). 'A computer program for quasi-continuous choropleth mapping', *Proceedings of the American Congress of Surveys and Mapping 48th Annual Meeting*, 1–3.

Fisher, H. T. (1966). 'SYMAP'. In Warntz, W. (ed.)(1970). *Selected Projects (1966–1970)* The Laboratory for Computer Graphics and Spatial Analysis, Harvard University.

Richer, S. (1976; revised 1978). *GIMMS Users Guide*. Department of Geography, Carleton University, Ottowa.

Tobler, W. R. (1973). 'Choropleth maps without class intervals', *Geographical Analysis*, **5/3**, 262–265.

Waugh, T. C. (1972). 'The Choropleth Mapping System—Reference Manual'. Inter-University/Research Councils Series *Report No. 7*. Program Library Unit, University of Edinburgh.

Waugh, T. C. and Taylor, D. R. F., (1976). 'GIMMS; an example of an operational system for computer cartography', *The Canadian Cartographer*, **13**(2), 158–166.

Waugh, T. C. (1977). 'A parameter driven language for use in application systems', *Proceedings of Symposium on Topological Data Structures*. The Laboratory for Computer Graphics and Spatial Analysis, Harvard University.

Waugh, T. C. (1979). 'GIMMS reference manual—2nd edition'. Inter-University/Research Council Series *Report No. 30*—second edition. Program Library Unit, University of Edinburgh.

Wolcott, H. M., and Hilsenrath, J. (1976). 'A contribution to computer typesetting techniques: tables of coordinates for Hershey's Repertory of Occidental Type Fonts and Graphic Symbols'. National Bureau of Standards *Special Publication No. 424*, U.S. Department of Standards, Washington, D.C.

Chapter 11
Future Research and Development in Computer-Assisted Cartography

D. P. BICKMORE

Attempts to forecast 'futures' are notably hazardous, and this is particularly so in a new subject that has not yet come of age. In such a case one cannot look back far enough in time to pick up much by way of perspective. Of course many of those who are already engaged in computer-assisted cartography (C.A.C.) are now working at the research and development end of the subject rather than in a major production sense—and their present research plans will inevitably extend some years into the future. Much of this is evident from preceding chapters of this collection of essays; indeed, the process of 'finishing' research can be notoriously long-drawn-out. But if that was all there was to it, my essay could stop here.

It seems to me that there are some broad preliminaries that need to be considered in this topic. The first of these stems from the confession that I believe cartography is no end in itself. I see it rather as an information science performing what is in essence a service function to a wide range of other disciplines. This should not inhibit cartography from having a research role of its own, but it does imply that the needs of those disciplines that cartography serves—to say nothing of the kind of data they furnish—will set the tempo and provide the raw materials for cartographic research. Those disciplines that are interested in spatial relationships—geography, geology, meteorology, astronomy, ecology, demography, to mention a few—will, I believe, always be the trendsetters for cartography. Hence it is probable that research in cartography—while distinct from research in, say, geology (Gold, Chapter 8)—demands the closest links with environmental science. If the main driving force in cartographic research is likely to come from this quarter where the need is, then the research cartographer needs a sharp awareness of the directions in which these sciences are moving and of those growing points in them which may call for the telling representation of patterns to help in understanding processes. So I assert the prime importance of the environmental disciplines to cartographic research.

My second preliminary must be to touch on the relationships with the range of computer-based technologies that the new cartography applies. On the one hand, I see this relationship as secondary to the link between cartography and the environmental sciences which clearly represents the *raison d'etre* of the subject. On the other hand, the computer and many of its associated technologies have arrived on the cartographic scene with such suddenness and with such potent muscle that they revolutionize the cartographer's approach and seem to transfer directly to his hand a new ability to analyse, or to extrapolate or to display, geographic ephemera or hypotheses in terms of a conversation piece, as Morrison has so eloquently argued in Chapter 2. This ability goes beyond the cartographer's traditional ability to provide the definitive end-product of a task conceived perhaps a decade earlier. The speed, the versatility, the questioning power of computer cartography seem to have dramatically increased the standing of the subject: cartography is now interesting to non-cartographers. Here again cartography is more than a special branch of computer science and it manages to identify its own particular contribution as something peculiar to itself and just as distinct from computer sciences and technologies as it is from the earth or environmental sciences. Perhaps the justification for any discipline lies in its ability to sustain a worthwhile independent existence; the independence of cartography is, I believe, an important standpoint in discussing the future of cartographic research and development: where does it lie, whence is it derived?

But having asserted that the computer-assisted cartographer is decidedly more cartographer than computer scientist, something must be said about the new potential that the computer can give the cartographer. Here one need perhaps do little more than reiterate that digital cartography is not just a new technique, as scribing was 30 years ago, for making the same old maps more quickly and more cheaply. The fact that spatial patterns can now be held in pure digital form means not only that they can be retrieved and made into visible maps at the press of a button, but that they can be at virtually any scale; on any sheet lines or projection; in virtually any graphical form; and, in principle, in combination with any other set of digital data. The process of going from the data bank to the display or to the plot also has a speed about it which is as remarkable as the versatility it so clearly brings. The change from the slow definitive map series printed in its thousands to the quick ephemeral display, uniquely generated *ad hoc*, must introduce into cartography a substantially new approach, even a new philosophy (cf. Chapters 2 and 3). So it is clear both that cartography already owes a substantial debt of gratitude to computer science, and that future research or developments within cartography will be heavily dependent on its links with computing.

Having talked of two of cartography's fraternal relationships—one old and fundamental; one new, excitingly powerful though expensive—it is now necessary to consider a third relationship: that of cartography and graphic design. This is a link that is as old as cartography itself, but one whose significance is even more evident as a result of C.A.C. than it had become in manual cartography. The new digital nature of cartographic data frees the language of cartography from its traditional analog forms, and in doing so provokes questions about the nature of the graphics that may be optimal for any specific cartographic purpose in communicating geographical patterns; obviously communication is a two-way business, with the map reader ideally able to read all the notes of the music and hence to see and understand the whole complex of spatial nuances that the map implies. Perhaps the cartographer's most important and difficult task is the ability to communicate pattern with minimum loss of accuracy and detail: no map can be judged as abstract art, and few that are of interest are also simple. Some dialogue does exist between cartographers and graphic designers (notably so in Switzerland) and some with perception psychologists. But to a formidable degree the cartographer (Robinson, 1978) relies in this field on his own experience and on the constraints of time, print, and design. These will be fortified in so far as he has a critical awareness of design techniques that seem to work in associated fields such as painting, typography, advertising, and film design. Neither environmental scientists nor computer experts are much at home in these design fields, but they have substantial relevance to cartography.

My purpose in discussing the three interdisciplinary relationships that affect the cartographer is to paint him as a jack of many trades, and to show by implication the need for this wide range of interests to permeate cartographic research and to show its distinctness from, for example, geology, computer graphics, and perception studies, but at the same time its dependence on keeping abreast of these disciplines.

Another preliminary which also needs consideration is to attempt some distinction, however hazy, between *Research* and *Development*. Other disciplines have perhaps achieved such a distinction with a crispness that is difficult to recognize in cartography. Some hold that research results directly from backing an individual; and that development is a more organizational—and expensive—activity. What is obvious, of course, is the interplay between research, with its idiosyncratic idea-generating capacity, and development with its ability to engineer useful results from bright ideas. One recognizes a kind of iteration between the two: thinking the unthinkable demands the longer and more arduous process of testing hypotheses, interpreting themes and demonstrating workability; and it seems often in these very development processes that further new ideas are sparked off. In C.A.C., however, the process demands the solid background of a system capable of responding to constructive new ideas. This often means specialized and expensive equipment and staff with a wide experience in using it. Under those circumstances it is all too easy for innovative research to subside gradually into a development phase where productivity becomes predominant.

It would be unrealistic to conclude my list of preliminaries without reference to finance and to the constraints that this inevitably—if properly—imposes on cartographic research. To a large degree, development work is constrained by the economic and social climate of the country in which it takes place. Research and development in computer cartography demands—and seems likely to continue to demand—not only expensive equipment, as has been outlined in earlier chapters, but also that sufficiency of time that will enable cartographers to address long-term research programmes as distinct from solving short-term and *ad hoc* problems. As such, it demands rich patrons—typically government organizations. And, of course, research work in cartography has to compete for funds with other research right across the board.

> Public attitudes to the benefits of scientific research have become more critical. There is (these days) a much more rigorous scrutiny of Research and Development programmes and of the structure within which they operate

so says a recent British Government report (*Review of the Framework for Government Research and Development*, 1979). Nor is it axiomatic that research in cartography will be favourably viewed by those taking note of forecasts which seem to point to continuing population increases and to the prospects of high unemployment and falling growth rates. In so far as C.A.C. is synonymous with labour saving its development is also likely to be resisted by organized labour. From all this it is evident that the economics of any case for cartographic research have to be persuasive, pointing realistically to quantifiable benefits in, for example, resource management. Experience shows that such cases are difficult to make and the momentum of projects that are established is not easy to maintain.

Having looked outside himself—at other disciplines, at the distinction between research and development, and at some severe economic constraints to research—the cartographer now perhaps needs to look inside his subject, to recapitulate what he really has to offer.

In these digital days perhaps the research cartographer needs first to reassert the essentially *graphic* nature of his subject; his concern has to be with geographical patterns presented pictorially in shapes and symbols, and perhaps in colours, and so as to be read in a different way—with a different part of the mind—to lines of text or columns of statistics. But the cartographer's use of graphics has at the same time to be a very disciplined one if he is to convey the maximum of information in concise form about the quantities or classes involved, and to do so painlessly as it were, so that deeper, more concentrated study of the map reveals increasing information and subtler relationships beneath the first—*gestalt*—look. For although the medium is a graphic one, the message is concerned with information often of a highly precise and quantified kind. Bertin (1973), in his *Semiologue Graphique*, has underlined that combination of artistic and scientific understanding that cartography involves: he compares the functions of the 'professional scribe' in the Middle Ages as a servant of a somewhat illiterate public, with that of the cartographer today. The computer-minded cartographer will do well to ponder this in considering the end-product of his researches, and to remember that a minority of any population is likely to find real difficulty in assimilating graphics (a deterrent, balanced perhaps by a minority with equivalent difficulties in assimilating tabulations). The cartographer's job has been compared with that of a painter; however, the painter has the great advantage of painting largely for himself and being little influenced about the meaning that his painting conveys to others. This is an indulgence that no cartographer can afford: his map is not just for himself, and he has to accept his service function as a constraint to his imagination. Communication is a strict discipline.

Cartography is a great correlator: a diagram really only becomes a map when it holds together two or more sets of geographical relationships in an explicit framework of grid or graticule. The play needs scenery—some base map detail has to set the scene for any foreground theme. Sometimes the difference between background and foreground will depend on the way in which a general map, such as a topographic one, is used at a particular moment; for example to read contours or to identify a route. One of the prime advantages that digital cartography can bring is the ability to select a background mix of features (and of the graphics by which they are displayed) in relation to the foreground subject of particular interest. This presents a potential escape from the traditional grey topographical base map which can obscure the main pattern with irrelevant noise. The research cartographer also needs a strong sense both of what topics illuminate and support one another, and of the graphics which will best display the relationship. In this traditional correlation exercise the cartographic designer can be enormously assisted by the computer with its ability to select from the data base the individual elements of the map, and to display them in whatever graphic form he may desire. As cartographic systems become more versatile they should provide scope for almost unlimited experiment in correlations, much of it producing ephemeral maps, sometimes as ends in themselves and sometimes as a preview of something more definitive.

An important aspect of the interrelation of different data sets in one map lies in the positional reliability of the data, either in its original form or in the classifications or groupings—quantitative or geographical—which the cartographer may use. The human eye has an ability to discriminate extremely fine detail, down to a tenth of a millimetre. This discriminatory power can give significance to positional relationships between different data sets, and can be—but rarely is—used with effect where correlation is between data sets with different positional reliabilities. Many topographic surveys have rules about the positional accuracy to which their data originally conform, but even these may vary as between the different kinds of features mapped—rivers, roads, contours, etc. Thematic mapping frequently calls for correlations between map elements with much more substantial differences in positional accuracy—for example, river patterns versus catchment areas, versus isohyets. These map correlations logically call for the employment of graphic symbols that give information about positional reliability in addition to their overt function of discriminating between the kinds of information they present. In so far as positional reliability has been an underrated element in cartography, computer methods appear to provide the means of holding reliability attributes for line or point data—the first step towards portraying them in the symbols used on the map.

The fact the photolithography has had a dominating technical influence on cartography during most of this century has raised the standing of coloured maps: the public seems to expect its road maps, its atlases, its geological maps to be printed in colour. For many cartographers the use of colour has represented a principal area of experiment and the photomechanical techniques of achieving it are an essential part of his professional background. One must, however, consider whether the emergence of automation in map-making will continue to be geared to professionally produced colour maps. The case against colour depends partly on its costliness—at least when the print run is small (for example thousands, as opposed to hundreds of thousands, of copies); furthermore, there exist all too many rather stereotyped 'committee designs' that make unimaginative use of colour in conforming to the disciplines imposed by the map series concept. In these two respects computer cartography should be able to make useful contributions: for example, colour television has demonstrated some dramatically beautiful effects, and as technology advances there seems no reason why this medium, in the form of Viewdata systems, should not be used increasingly by cartographers (probably more as a supplement than a substitute for printed maps) and even larger audiences (*Financial Times*, 1979). Rather the same technology may well provide colour C.R.T. displays at the end of computer networks, and hence provide the means of colour digital maps as between relatively small research groups, planners, etc. Some more considerable questions are whether the ability to use a large number (for example, 250 or more) of different colours and to provide rapid changes of colour will be a helpful as well as a novel element in cartography; whether using a map on a display will permit using the same fineness of minute detail that is traditional in cartography (perhaps it will make it unnecessary, since 'zooming in' may be found preferable to 'peering at'); and whether this kind of colour display will lead to an effective dynamic cartography. Economy is patently one of the prime constituents of good design, and the cartographer's need for colour depends, as ever, on the complexity of the pattern he is presenting. One of the most substantial assets of computer cartography—based on a versatile data bank—is the ease and economy with which data can be grouped and regrouped, using different symbols, or scales, or map areas, or different mathematical approximations. In such a setting, the cartographer's scope for creative design is likely to be vastly enhanced, and he will find computer-generated colour as one among an increasing number of the tools of his trade.

Cartography is obviously concerned with displaying, analysing, and integrating patterns, but it has so far made little attempt to handle process, and has barely come to grips with time as an additional cartographic dimension. Information about process as well as about pattern is often desirable in environmental science, and data with both time and space elements are becoming available. Some such data—typically in the field of meteorology—are directly collected in that way; other data sets can lend themselves to being arranged so as to provide a time element—successive population censuses can provide some indications of movement; other data sets can be worked on by computer modelling processes to provide forecasts of future patterns. 'Dynamic cartography' is an attractive concept making use of the difference between the 'still' and the 'movie' in photographic terms. However, early attempts to use computer filming have run into user difficulties, and high costs have not always equated themselves with commensurate increases in scientific understanding. In addition, the paraphernalia implicit in the film audience situation seems an over-elaborate way of map reading—a conducted tour as opposed to a browse. These difficulties re-emphasize the need for cartographic design work based on sympathetic understanding of the environmental problem, of technical developments (especially of versatile data base structures), and of access to the necessary and costly equipment.

The research cartographer will also need to consider how much of his future data will come to him in digital form. This will sometimes be for the good and sufficient reason that the data were originally collected in that form. A substantial and famous example of digital data collection is Landsat whose digital tapes seem now widely accepted as the optimal form—at least by those with computer graphic facilities. But this same tendency is reflected in more and more subjects—geophysics, hydrology, oceanography, to name only three. In some of these cases the data are not continuous—as for Landsat—but collected at sampling points, and subsequent computer processing is used to rationalize predictions of what happens between sampling points. Such data will come for mapping in digital form regardless of whether the original sampled data were digital or not. The more the cartographer can be relieved of the tedious and error-prone business of converting raw data into digital form, the happier he is likely to be, and the more he can concentrate on the design elements of his work. Nevertheless, a preponderance of his data will—perhaps for the rest of the century—be in the form of existing maps or patterns and hence likely to demand digitizing if it is to be effectively tied in with new specialized data. (The costs and problems of digitizing have indeed overshadowed computer cartography during the 1970s but, as Boyle has indicated in Chapter 4, a range of automatic scanning processes may soon come to the rescue—to be used in conjunction with interactive graphics for converting existing maps into computer form; but this will all take time and money.) So it is likely that much specialized thematic data will be ready-made for computer mapping while many of the necessary base maps will not.

The assumption that specialized data in digital form can be overprinted on existing non-digital base maps—whether they be topographical, geological, land use or whatever—is one that needs some care. In a simple sense, of course, they can. Thus on large scales (for example, 1/1250 or 1/2500 in Britain) where positional accuracy has not suffered by generalization, the new lines, points and areas of special overprints can be placed with precision in relation to existing lines on the base map. Difficulties, however, arise with smaller scales where the traditional base map may have been manually generalized and its shapes and positions minutely altered. In these conditions (which are not infrequent as a result of generalizations quite properly done for their original purpose) accurately located new data may find themselves out of tune with the base; for example on the wrong side of a river or road. But traditional methods of map generalizations do not account for all misfits between data sets, and it is my guess that the cartographer of the future will find that his ideal new data sets in digital form may still contrive to pose problems of local compatibility with the resultant need to edit and adjust in finicky detail familiar to centuries of cartographers. Such editing processes remain expensive, even by computer; they may remain a practical constraint for some years to come, and one that seems endemic to the subject.

Cartography has always been obsessed by problems of updating, and one of the principal claims of computer-assisted cartography is that it will simplify and economize that process. That claim which seems likely to be substantiated, or refuted, in the 1980s is, of course, based on unique data records from which secondary, tertiary (or whatever) data sets can be directly and very rapidly derived. The principle of unique records leads of course to the assumption of a single data base from which all scales of maps can be derived. Practical complications inherent in this assumption are highly germane to cartography (and they are addressed below) but the benefits by way of updating at least seem significant even if they emphasize a definitive centrality in the data base in contrast to the peripheral and indeed the ephemeral nature of the resultant maps. For many map-makers this sacrifice of the definitive map seems a formidable price to pay.

So much for idiosyncratic views of the cartographer's relationship with cartography. The title of this chapter does, however, demand more than this and implies that it may be possible to identify distinctive research fields for cartographers of the future to address. I believe two such fields are of prime importance: (a) the cartographic data base; and (b) cartographic design based on the new flexibility the data base gives.

The Cartographic Data Base

The comparison between a library and a data base seems a natural starting point for brief discussion of this subject, if only because it illuminates some important, if elementary, differences. Thus the unit in a map library is the particular map, whereas the unit in a cartographic data base is the unique map element as represented by points, or line segments (which linked together form areas). The points or lines will, of course, carry attributes which define their description rather than their locational function, for example, as rivers, administrative boundaries, fault lines. And there will, generally speaking, be many attributes for each point or line on the assumption that the positional information about the point or line needs to be unique (if a boundary and a river are identical in position there will, in the data base, be at least two attributes but only one line segment). Thus the structure of the data base is fundamentally different from the map library in so far as the one is directly related to information and the other indirectly and via maps. These basic differences seem to go a good deal further. Thus a map library is necessarily organized in terms of the sheet line system of individual map series or (ideally) into the pages of individual atlas maps. The data base by contrast makes the assumption of a continuous carpet of cover of the entire area concerned and, in doing so, seeks to provide for the retrieval of any required rectangle—or polygon—within that area so that the map user can arrange his own map edges to suit his area of interest on any particular occasion. Considerations of map scale under a data base system lose much of the inflexibility they have in the traditional setting—this partly comes from the need to have more flexibility in retrieval (in television terms, zooming-in). Scale also seems to take on something of a secondary importance in relation to completeness of cover: this implies that the cartographic data base will often be made up as a patchwork of scales depending on what is available, what is most up to date, what has been digitized, etc. Scale differences within the data base will appear as differences of resolution—and hence of positional reliability—since scale itself does not become meaningful until a map is actually plotted. It seems to follow from this assumption that the cartographic data base will have some inherent untidiness arising from varying scales and consequential edge-match, or patch-matching, problems, in line positions, and in differences of content and classification between original maps whose digitizing has created the data base. Elegant solutions to these problems both in topographical and thematic data bases will call for considerable cartographic skills.

Another comparison may also be illuminating—this time between cartographic data bases and national or regional atlases whose production *de novo* or in terms of new editions currently seems active in some fifty countries of the world. National atlases are, of course, cartographic attempts to display inventories of resources in the geographical sense. As such they suffer in so far as they freeze information at a particular time—and expensively so—and are outdated by subsequent information; for example a latter census, or a new patch of survey. But despite these shortcomings, enthusiasm for making such atlases patently remains. They also suffer from problems of comparability in so far as each map or set of maps on distinct topics is likely to have had its origins in quite diverse data (as between demographic characteristics and land-use patterns, for example) collected independently and for different purposes. Cartographic data bases have many of the same objectives and seem to offer better prospects of updating if not necessarily of assimilating data from sources which are different if not actually incompatible. In theory, the data base approach provides for an infinity of correlations—as the user wishes. Doubtless, the analogy of the national or regional atlas serves well for those concerned with data bases. Doubtless, too, the prospect of using computer networking will bring the ability for a particular specialist with his own data set to access background data from another discipline with ease and rapidity, as and when he requires it, and in such a form as to join the two together cartographically. The practical achievement of such aims seems a proper field for applied cartographic research closely allied with computer technology—a new kind of regional geography.

If cartographic data bases have some elements of the map library and share common aims with national and regional atlases, they will inevitably be large affairs and their size and complexity seems likely to impose some strain on computer science. It has been asserted that computer architecture has been more concerned with solving differential equations than with handling the inter-relationships of vast quantities of information—relationships which go far beyond simple hierarchical structures, as for example Bouillé's work (1978) on hypergraphs has shown. There is also evidence that small pilot exercises have a way of concealing problems that only materialize under full-scale working conditions. These matters fall properly in the field of computer science but none can doubt that their solution is a basic act of faith that the computer-oriented cartographer has to assert. He can, however, make some contributions to the issue.

The first contribution lies in defining speed of operation—that is to say how quickly the map user needs his map. Here the acid test may be an economic one— how much more will he pay to have his map in 10 seconds rather than in 10 minutes; overnight rather than next week? A planning map must be ready for a meeting next week, while a weather map may be irrelevant next week; a contour plot of bathymetric data may be required before the survey ship leaves the area; a telephone or Viewdata enquiry may be pointless unless answered in seconds. (The response speed may, of course, affect the graphics but the first graphic step is to find and retrieve the data.)

The second contribution relates to the topics that may be in demand: some base map data may be standardized and held, as it were, ready-made in graphical form, but in doing so it runs the risk of being out of date with the primary data base.

His third contribution may lie in the field of defining whether the data base is slimmed down to the barest data essentials from which, for example, contouring programs will develop sets of derived grids which in their turn will yield cartographic patterns. There is a trade-off between the advantages of holding only the bare essentials and the costs and time of such intermediate processing (which it may be unnecessary to perform repeatedly for different applications). Rather similar questions arise in terms of boundary information which may have its 'purest' form as unique line segments with multiple feature codes defining attributes either side of the boundary (or of centroids linked to the line segments which provide this information). Data base economy suggests that only the unique line segments and their feature codes be held: retrieval arguments suggest duplicating common line segments so that complete files can be held for each areal unit with a set of all relevant boundaries (and arranged in clockwise or anti-clockwise sequence). In this way areal calculations or colour plates can be retrieved more rapidly. These are matters where cartographic judgement seems to have an extremely important function.

A fourth contribution by cartographers lies in establishing sets of standards oragreements for data that are to be accepted by data banks. Some of these standards will relate to digitizing accuracies and will need to take note of the positional accuracies of the data as collected; others will relate to data compatibility and to such elementary but essential issues as ensuring, for instance, that a geological boundary that goes to a coastline picks up a common topographic boundary for that feature rather than introducing a different duplicate boundary of its own. Obviously, common data formats are as desirable as they are difficult to achieve in a human world.

Cartographic management will identify other controls that it believes desirable— the attribution of costs, the scheduling of updates, copyright issues, security issues, relating both to the data and to the possible corruption of tapes or disks. The new scope for cartographic standardization can, however, all too easily proliferate, and in doing so it may dangerously circumscribe a developing subject and its new-found flexibility. This danger seems a particularly significant one in the growth of cartographic data bases.

These kinds of issues will, doubtless, become organized and refined into agreed practices as cartographic research and development proceeds. At the same time other disciplines (Scarrott, 1979) will be concerning themselves with the development of searching machines, the pros and cons of vector or raster systems, the use of holographic stores, the manifold applications of such things as array processors; in keeping track of these and endless other exciting possibilities, the cartographer may need to remember two points first, that cartography is a minority issue in the computing science field, both in terms of money and of interest; and second, that cartography will be judged more by maps than by methods.

Cartographic Design from the Data Base

The term 'realization' in the context of C.A.C. serves to emphasize the difference between a massive digital mountain of information quite devoid of particular graphic implication on the one hand, and its translation into usable, visual forms on the other hand. And it is, of course, essential to the graphic freedom inherent in the new methods that the data base should not trespass into the graphics, nor the graphics into the data base. They should complement each other; on the design side I believe cartographers can claim the dominance which computer scientists can assert on the data base side.

In using the data base the cartographic designer will need to assimilate two admonitions: first that there is no merit in computer maps *per se*; and second, that the arguments for attempting to mimic existing map or chart design needs handling with suspicion if not with scepticism. That said, let us consider the potential of the new cartography.

Perhaps the basic element lies in the speed with which it is possible, under computer systems, to move data to graphics—a concept idealized as 'the instant map'. Developing even this capacity to its full potential is still limited today; thus, for example, in the production of coloured geological maps by computer there still remains, and for valid economic reasons, a substantial element of traditional photo-lithographic hand work; to this may be added time delays of printing and binding, and these accumulate to an extent that seems to dwarf the rapidity with which a particular set of geological symbols and boundaries can be derived automatically. It is unfortunately true that the cartographer has to concern himself with *total* elapse time from calling up data from the data base to placing the hard copy in the hands of those who need it. Perhaps the identification of new methods of distributing maps, for example via Viewdata or similar methods, may turn out to be as important as the present concerns with fast displays or plots within the computer laboratory. At present, however, over 90 per cent of map use seems to be in situations demanding hard copy and there is a long way to go even before networking systems as between computer laboratories become practical realizations.

Another basic constituent lies in the ability of the computer-generated map to give a direct reflection of primary data, quite undistorted by intermediate, secondary, tertiary, etc. representation. The term 'data warehousing' implies a stock which may take many forms, such as the results of economic or social surveys, or of meteorological data; in such cases direct data reduction and retrieval from source is highly prized: so too in cartography. This direct route to source data produces its problems, not only in the structure of the cartographic data base but also in graphic designs: many of these problems go under the name of generalization, a term that encompasses several elements. Thus, consideration of class intervals (Evans, 1977) is an abiding cartographic concern, deriving from the attributes of the original data. For example, the linear network of a road system may be grouped into administrative classes (A, B, or C roads) that imply minimum engineering capabilities, but at source some of these capabilities such as width or curvature or surface, may be described in more detail, while other attributes are also associated with the network, such as traffic flows. In making the very proper claim to have access to increasing multiplicity of source data, the cartographer has equally to have an equivalently increasing repertoire of designs by which to translate the new wealth of data into readable maps. Many of these designs have to assume that noticeable differences between symbols, for example in line weights or decoration colour, can be equated with significant steps in the source data. At the same time as such quantified symbols for a particular topic yield, on inspection, increasing information, so also the symbols will need to be seen in series as of a generic graphic type and distinct from other elements on the map. Social, demographic, medical, and industrial geography have traditionally produced many such problems but C.A.C. seems to demand new graphic skills if representation is not to be a limiting factor on the vast data potential that is now at call in the physical and social sciences.

Another aspect of generalization that applies both to topographic and thematic cartography has to do with the handling of shape under conditions of substantial reductions of scale. Over the years cartographers have found traditional solutions to some of these problems largely on the basis of using a relatively limited number of different graphic symbols—the double line for a road, the open circle symbol for a village or town. Often these symbols have served well enough, for the user easily accepts them as familiar and, under traditional circumstances, he has no easy means of overlaying additional cartographic information that may produce inconsistencies in positioning. But under new circumstances where there is a direct link to source data, and to position in particular, there is everything to be said for preserving a rigorous positional accuracy overall and, hence, relative positional reliability between different kinds of data. While the retention of positional accuracies throughout changes of scale would seem to be an obvious tenet of cartographic faith, it is surprising how frequently adjustments to position are made by cartographers. Their justification seems based on adherence to a particular and limited set of conventional signs. Of these, the double line road symbol is a good example somewhat hallowed by a century of use: it does, however, represent at smaller scales a swathe of country several times wider than the space occupied by the road itself. Consequent positional shifts may therefore be required as between those line features which are contiguous such as rivers or railways or land-use boundaries and which run more or less parallel to each other, or between line and point features such as groups of buildings whose positions may be obliterated or distorted by the swathe of the road symbol and so also demand repositioning. The acceptance of the need for such positional shifts introduces the need for a pecking order, stating which feature is the more plastic and mobile and which should be accorded a measure of priority in retaining true position. It seems extremely desirable that digital cartography with its direct access to source position should avoid these complicated and continual editing processes of repositioning, by a studious re-examination of those symbols, colours, etc. which may be able more exactly to represent natural pattern. And already, the multiplicity of satellite pictures now available demonstrates quite incidentally the legibility of, let us say, a secondary road a 1:1 million scale and in doing so seems to imply that line legibility is not directly related to width or weight of symbol but derives in part from its own linearity.

And what seems true of linear generalization seems true also for area generalization. A large-scale or indeed a data base source may indicate an archipelago-like situation, not necessarily only of land/sea but of woodland/farm land or gravel/boulder clay or built-up/rural areas. Small-scale traditional representation of such patterns generally makes a graphic assumption of a minimum 'island of colour' (often of around 4 mm^2) surrounded by a black pattern line. Under such conditions smaller 'islands' are omitted and by implication contribute their areas to the surrounding 'sea'. Solutions of this kind are common in most thematic mapping and are defended, and at the same time complicated, by practitioners who claim to interpret their own 'representative islands' so that the overall pattern is in some artistic sense preserved despite departures from observed detail. Again, comparison with air photography or satellite imagery is illuminating in that it preserves natural pattern right down to the resolution of the original data. In doing so it dispenses with the device of the black pattern lines, and hence permits a great decrease in the size of minimum islands of colour. These design differences may well provoke thought among research cartographers whose links to raw data are faster and more direct and whose control over scale, colour, and symbolization are now potentially far greater. Furthermore, in a given time, the computer cartographer can offer a dozen and more experimental alternatives to his predecessors' single solution.

Another basic difference in cartographic design of the future concerns the representation of the same geographical information on many different forms of graphic device—from high resolution flat bed or drum plotters operating up to one thousandth of a millimetre, to plotters or even lineprinters functioning to a resolution many orders of magnitude lower. The benefits that may seem to flow from being able to distribute map data far and wide in digital form seem to call for examples of the display of identical data sets on quite different graphic devices. It seems possible, for example, that shortcomings in graphical elegance on a low-resolution system might be compensated by the ability to produce more separate analytical maps for the same cost as by a single more complex synthesizing map from a high-resolution plotter. The revolution in the use of grid squares for mapping that followed the introduction in the mid-1960s of the SYMAP programs has been remarkable in the history of cartographic design: it seems to postulate that there will be other far-reaching design solutions in this field, not necessarily in terms of grid squares.

Earlier in this chapter attention has been drawn to fields in which cartographic design has not so far contributed in any major way in contrast, for example, to the development of contours in the nineteenth century and to SYMAP in the twentieth century. Four such fields seem to be worth identifying here:

(i) mapping of movement and the use of time dimension;
(ii) the explicit handling of variations in positional reliability within one map;
(iii) the handling of multi-dimensional quantitative data on a single map;
(iv) the representation of patterns within a solid (whether, for example, for geology, oceanography, or town planning).

Satisfactory solutions to such problems not only lie in experiment with symbols and colours, linked to the clever use of software or hardware systems; they also depend both on getting the map to the user faster, and on persuading the user to take the mental effort to assimilate new symbols and to use what may even amount to a new cartographic language. The growth in the size of the data base, its rapid accessibility, and the wealth of quantified detail on offer, make new calls for the cartographer to match in his design the versatility and inventiveness of his colleagues in the environmental sciences and in computer technology.

If ever there was a time for map design to come alive it should surely be now, when we seem to be at the end of the beginning of the new cartography (Margerison, 1976).

REFERENCES

Bertin, J. (1973). *Seminologie Graphique*, 2nd edn. Gauthiers-Villars, Paris.
Bouillé, F. (1978). *Hypergraphs and Cartographic Data Structure: the HBDS System*, International Cartographic Association, Maryland.
Evans, I. (1977). *The Selection of Class Intervals*. IBG New Series, vol. 2, Contemporary Cartography.
Financial Times Survey (9 May 1979). *Viewdata Systems*.
Margerison, T. A. (1976). *Computers and the Renaissance of Cartography*. National Environment Research Council, London.
Review of the Framework for Government Research and Development (1979). (Cmnd 5046). HMSO, London.

Robinson, Arthur H. (1978). *Programme of Research to Aid Cartographic Design.* International Cartographic Association Commission on Communication in Cartography, University of Wisconsin.

Scarrott, G. G. (1979). *From Slave to Servant: the Evolution of Computing Systems.* Clifford Patterson Lecture, Royal Society, London.

Index

Accuracy, 7, 82
  of machine production, 19
Animated cartography, 11, 22
ASCII, 65
Audio input, 45
Automated cartography,
  definition of, 1, 2
Automated mapping, definition of, 2
Automatic line followers, 40
Automatic line scan digitization, 42–44

Bar shading, 223, 231

CALFORM, 205
Canadian Agricultural Census, 29
Canadian Federal Surveys and
  Mapping Branch, 77, 79, 90
Cartographic,
  analysis, 67
  design, 245–48
Cartography, 4, 10, 11, 60
Census,
  mapping, 191
  research unit, 29
CGIS, 44
Choropleth map, 22, 30
Classifications, 10, 16, 22
Cluster analysis, 164
COLMAP, 214
Colour maps, 239
Compilation, 14–16
Computer-assisted cartography, 1–2, 25–37, 39, 84, 91, 93, 192, 235–37
Computer mapping, 2
Conformal map, 61
Coordinate system, 15
Costs, 3, 9, 55–56, 88
Crosshatch, 223, 231
C.R.T.s, 19–20, 39, 240
CYCLOPS, 34

Data,
  base, 2, 15, 32–35, 89–91, 183, 242–48

capture, 12–13, 39, 181–87
manipulation processing, 68–69, 74–75, 86–87
smoothing storage, 13–14
structure, 11, 13–14, 22
Definitions, 10, 22
Design, 21–22
DIDS, 207
Digital,
  mapping, 28, 149
  topographic data, 29
Digitizing, 13, 22, 40–42, 71–75, 85, 106, 125–32
Discriminating analysis, 164
Display screens, 52, 67, 77
Dot,
  maps, 113
  screen, 223, 231
Drafting, 54–55, 76

Editing, 14, 17, 133
Education, 20, 22
Enhanced image, 19, 20, 22
European Economic Community, 29
Expanding capabilities, 11

Fabric analysis, 167
Facies map, 162–66
Factor analysis, 17, 164
Founts, 18

Geocartographic,
  objectives, 222–23
  system, 193, 222
Geochemical maps, 158–62
Geocoded, 193, 201
Geological maps, 169–76
Geophysical maps, 155
GFB/DIME, 194
GIMMS, 18, 205, 207, 219–34
GPIS, 225, 227
Graphic design, 236
Graphics, 27

251

GRDSR,	196, 213	PILLAR,	205
		Plotting,	139
HBDS,	173	PREVU,	204
Hershey alphabets,	230	Printed map,	19, 21
		PRISM,	205
IDAK,	95	Production efficiency,	9–10
Interactive display and edit,	48–52	Proximal maps,	144, 205
International Cartographic Association,	1, 4, 28	Raster,	
Isarithmic maps,	112–13	data,	12–14
Isoline maps,	140, 145	scanning,	40, 43, 77, 91
		Replication,	10, 16, 19
KASQF,	96		
		Scanning,	42–45
Laser plotter,	19	Semivariogram,	162
Leda system,	183	Shaded maps,	114
Lettering,	18–19, 21	Simple maps,	21
LINMAP,	214	Simplification,	16–17, 22
Lithological maps,	169–76	SIMULA-67,	173
		Soil maps,	124
Machine production,	15, 18–20	Soil science,	123
Machine resolution,	7	Speed of production,	7
Manual digitizing tables,	40–42	Stream data,	12–13
Manual production,	9, 11	SYMAP,	27, 204, 207, 248
Map libraries,	62	Symbol placement,	17–18
Map user community,	6, 21	Symbolization,	16–18
MAPMAKR,	196		
Methodological change,	10, 12–20, 22	Temporary map,	7, 11, 18–22
Multivariate maps,	144	Thematic data,	15
		Thematic mapping,	108
National Land Survey of Sweden,	94	Topographic maps,	60, 61, 176–81
Natural Environmental Research Council, Experimental Cartography Unit,	28	Topography,	60
Ordnance Survey (UK),	61, 78	US Bureau of Census,	29, 194
Orthomorphic,	61	US Federal Agencies,	20
		US Geological Survey,	29
Perceptual limitations,	9, 14		
Philosophy of cartography,	12	VERSATEK,	19
Photogrammetric,		VIAK,	96
compilation,	80–81	Virtual map,	19, 20, 22
input,	46		
Phototype setting,	104	WISMAP,	205